Anonymous, Hodges Foster

Lectures on Public Health

Delivered in the Lecture-Hall of the Royal Dublin Society

Anonymous, Hodges Foster

Lectures on Public Health
Delivered in the Lecture-Hall of the Royal Dublin Society

ISBN/EAN: 9783743395411

Manufactured in Europe, USA, Canada, Australia, Japa

Cover: Foto ©berggeist007 / pixelio.de

Manufactured and distributed by brebook publishing software (www.brebook.com)

Anonymous, Hodges Foster

Lectures on Public Health

LECTURES ON PUBLIC HEALTH

DELIVERED IN THE

LECTURE-HALL

OF THE

ROYAL DUBLIN SOCIETY,

DUBLIN:

PUBLISHED BY

HODGES, FOSTER, & CO., 104, GRAFTON STREET.

————

1874

DUBLIN: PRINTED BY ALEXANDER THOM, 87 & 88, ABBEY-STREET.

FOR HER MAJESTY'S STATIONERY OFFICE.

THE lectures which are comprised within the present volume were delivered in the Lecture Hall of the Royal Dublin Society, in the months of April and May, 1873, as the fourth annual series of Afternoon Scientific Lectures.

Although the Council are bound to guard the Society from being held responsible for any of the opinions advanced by the several lecturers, yet they feel assured that the eminence of the gentlemen who so generously responded to their invitation by devoting a large portion of their time and thoughts in the discussion of the important public questions which form the subjects of the lectures, cannot fail to secure for each the most respectful and attentive hearing.

The Committee of Science of the Society, with whom the design of the series originated, were fortunately aided in completing the arrangement for the delivery of the discourses by the Council of the Dublin Sanitary Association, to which latter body the Committee desire thus publicly to express their thanks.

MAY, 1874.

CONTENTS.

LECTURES ON PUBLIC HEALTH.

LECTURE I.

ON SANITARY SCIENCE IN IRELAND.

By WILLIAM STOKES, M.D., D.C.L., F.R.S.,

Regius Professor of Physic in the University of Dublin.

LADIES and GENTLEMEN,—I have to crave your indulgence for the many shortcomings that must appear in this lecture, and I trust you will remember the comprehensive nature of the subject, and understand that it is impossible to give more than a chalk outline of it, bringing into relief such points as relate especially to the health of this city.

Sanitary science, which embraces everything that can prevent disease, and so promote the public health, has long occupied the minds of thinking men in other countries, and now in Ireland the question is asked, what is the meaning of the term, so familiar a one in England, on the Continent, and in America? Some of our fellow-citizens, lay and professional, have established a Sanitary Association in this city, and the Science Committee of the Royal Dublin Society, acting in concert with the Association, have resolved that lectures upon the subject should be delivered in this theatre. This augurs well for the future. I may say that these lectures will not be for the purpose of announcing any novel doctrine. Their object will be to draw the attention of our fellow-citizens to the nature, the great importance, and the extent of the subject.

In this attempt, which is not the first of the kind in Dublin, we feel ourselves at a disadvantage. The subject is a prosaic one, and the public mind has too long been travelling in another direction to remember that Ireland has been frequently the source of epidemic disease, and too long heedless of that sanitary reform which has grown all round her, and the adoption of which is so important to the material

B

prosperity of the country. In relation to our subject we have nothing to stimulate popular excitement. We have no subjects for oratory, even if we had the power to use it. But we are here to tell you that which we know, and which must be known and thought upon, and to bid the victims of misery, physical degradation, and the unsparing pestilence, plead for themselves—

> " Show you dead Cæsar's wounds,
> Poor, poor dumb mouths,
> And bid them speak for me."

As regards public sanitation, our English brethren are far ahead of us ; not very long since they were in much the same condition as ourselves, and had to combat ignorance, and its offspring, ridicule. But in England that stage of the question has been passed, and we may be satisfied that by patience, diligence, the exercise of sound judgment, and the eschewing of theory, we may promote the cause in this country. It is many years ago since Dr. Rumsey, of Cheltenham, at a private meeting of friends, gave the first impulse to a movement which is now everywhere felt in England, and has attained such large proportions that a.Royal Sanitary Commission was appointed to examine into the public health, and also to investigate the state of the sanitary law. On the report of the Commission important Acts of Parliament have been introduced by the Government, and the great step taken of the creation and appointment of a Minister of Health —a central authority for regulating and controlling all the laws relating to the physical well-being of the community. A moment's thought will show the great magnitude of the subject. The sanitary laws on the statute-book were multifarious, often anomalous, or even contradictory. Consolidation and revision were necessary to form a comprehensive and intelligible sanitary code, the provisions of which it is hoped will be in accordance with the actual state of science, and the social condition of the people. It gives me pride to say that, indirectly at all events, and in one respect directly, Dublin has borne its share in the good work. It was at the annual meeting of the British Medical Association, held, on the invitation of the heads of the University, in Trinity College in 1867, that it was resolved to petition for the appointment of the Royal Commission of Inquiry. In this the Social Science Congress joined, and shortly afterwards a qualification of a high order in State or Preventive Medicine was established by the Board of Trinity College, the only one of the kind as yet existing in these countries.

Now let us specify some of the subjects with which sanitary legislation must deal. Before enumerating them we may glance at two great points which should guide or influence a Minister of Health. These are indicated by Dr. Rumsey. The first is the creation of a fitting machinery for investigation, and the second the preserving a congruity in the various laws that from time to time may be enacted—congruous they certainly should be, as they are to subserve a common purpose and be worked by a central authority. But whether we can hope for such a complete and comprehensive project as that which Dr. Rumsey has shadowed forth, wherein, to use his words, " each part shall at once and harmoniously fit together like the details of some vast pile which, when completed, would exhibit a unity of design and stand in simple grandeur," is a question. We must not forget that our actual knowledge is still imperfect, science being ever progressive and ahead of us. The subject of public sanitation, which implies Preventive Medicine in its widest sense as distinguished from Curative Medicine, touches every hearth and home in the country; every man, woman, or child from the highest to the lowest; every institution in the State, its power, its defences, its education, its manufactures; every trade, every occupation, domestic purity, domestic happiness, national prosperity, national health, longevity, and morals; the duties of property, the exercise of charity, and the blossoming and the fruit of our common Christianity. Its end is to improve and to preserve man's body in the best condition, and through it his immortal part. " The body of man," says Dr. Acland, " is not only the casket which contains the soul. It is more—it is a casket which, under certain conditions, moulds and modifies the soul." He says again, " There is a class of persons, I am happy to say rapidly diminishing in number, who seem to be of opinion that we who are engaged in sanitary work are somewhat fanatical; and that because it is connected with our material frame, it is therefore a second-rate subject, fit only for inferior men." Further on he says, " If I can read anything in the history of the globe it is this—that the great qualities of a people depend in large measure (except in rare instances) upon the physique of the nation. I appeal to all—that is, those who have studied the philosophy of history—whether it is not the fact that some of the highest of human qualities have been shown in a most eminent degree in days when there was a noble physique, but when there was no systematic education, according to the modern notion of book learning." He adds, " that so far as the comparative national health is concerned, I say there is no possibility of exaggerating the importance, not to our own country alone, but to the world, of fostering and caring for the body of man."

There are so many important subjects relating to State Medicine, that the first requirement in establishing that machinery for investigation is the education of a class of men able to judge and report on all questions relating to public health. I do not like the term "expert," which implies excellence in but one or two subjects. But State Medicine has a wide scope, and the adviser in it must, for purposes of Government, have proportional powers. Public experts in certain departments there must be, but the adviser as to public health should have a natural aptitude for the subject. He must be highly and generally educated. He must know all the conditions under which the health, vigour, and life of man are best preserved. It is well that he should know the main facts of Curative, as contrasted with Preventive Medicine; be familiar with vital and sanitary statistics, and, by analysing and classifying the results, be enabled to draw just conclusions from them. Such a class of men is now being formed, and the community will demand that the knowledge which they profess will have to be represented in every governing body in the country which has to deal with the physical well-being of the population, from the Imperial Government to every local administration. The status and the rewards of such a class will be measured by the importance of their duties.

Upon the registration of causes of death the public mind is in need of some enlightenment. It requires to be shown that to determine the cause of death is often one of the most difficult of problems. In most cases the discovery after death of an anatomical change is held to account for the death—so it appears on the registry. But such a conclusion may be, and too often is, utterly erroneous. As a natural organic structure throws but little light upon life, so may it be said that diseased structure fails, except in a very few cases, to explain death. In the so-called zymotic diseases the anatomist can say only what they are not. Every day the mistake is made of confounding the effect with the cause. Men die with local disease, but not necessarily of it, and it may be held that in a vast proportion of the deaths of the eight hundred millions of the world, the immediate cause of death remains hidden from us.

In reference to the important question of the registration of sickness, it is plain that there is abundant opportunity for doing it, if not perfectly, at least effectively, by the State registration of sickness in every institution supported by public funds. Dr. Rumsey observes, " that there are literally no published records of the cases of sickness attended at the cost of the community. The sanitary state of the people is therefore inferred solely from the number of deaths—that is, from only one of the results of sickness—no public account being taken of the number and duration of the attacks which shorten the effective lifetime of the population." " Facts are accumulating to prove that the mere number of deaths occurring in any locality bears no constant or even approximate ratio to the real amount of unhealthiness existing there. As a necessary result of improvements in domestic management and medical treatment, and owing to the removal or absence of those more virulent agents of destruction which prematurely sever the threads of life, its duration has been lengthened in our great cities, but at the same time the sickly and infirm period of existence has been prolonged probably in a greater degree than even life itself. Chronic diseases, or, at least, functional disorders, have increased, vital force is lowered, man's work is arrested, his duties are unperformed, his objects fail, though he still lives. Weakly, diseased children are now mercifully helped, as they never were in olden time, to grow up into weakly, ailing adults, whose children inherit their unsoundness. Is this true sanitary progress ? Does it deserve the ostentatious parade of a decreasing death-rate ?" Lastly, personal antecedents and remote causes of death now generally escape notice. Dr. Rumsey further shows " that the deaths of those who enter a district merely to die there belong rightfully to another locality. This applies to large towns, seaports, hospitals, workhouses, and so on. The mere death-rate, therefore, as a measure of local unhealthiness, leads to most fallacious conclusions." While I agree with Dr. Rumsey in these statements I am far from denying that an accurate knowledge of the death-rate is of the greatest importance, especially in the absence of trustworthy statistics of disease.

To those who have not thought upon the subject I may read a list of details which must pass under the control of the Minister of Health. It is given in an address to the Social Science Congress, by Dr. Acland, and as the speaker observed that the list is not intended to be exhaustive, I may add a few particulars to it. I shall have to bespeak your patience, but I know that I address an enlightened audience that will bear with me. The first subject is that of water supply ; 2nd, removal of refuse ; 3rd, control

of buildings, new and old; 4th, sewage; 5th, drainage; 6th,
prevention of over-crowding; 7th, prevention of contagious
and infectious diseases; 8th, prevention of epidemics; 9th,
formation of sanitary areas; 10th, registration of births, and
of causes of death; 11th, registration of sickness, as
distinguished from death; 12th, the laws of quarantine;
13th, the superintendence of epizootic epidemics; 14th, the
inspection of all hospitals, including those of the insane;
15th, the sanitary inspection of penitentiaries and prisons;
16th, the inspection of dispensaries; 17th, the inspection of
druggists' establishments; 18th, the inspection of factories;
19th, the inspection of the dwellings of agricultural labourers;
20th, the placing of every district within reach of a sanitary
inspection by the best scientific and medical experts; 21st,
education and registration of medical practitioners; 22nd,
education and registration of nurses and midwives; 24th,
control of intoxicating liquors; 25th, mode of obtaining
analysis of air, food, water, &c.; 26th, vaccination; 27th,
coroners; 28th, vested rights of mill owners and others to
interfere with sanitary measures in new populations; 29th,
organization of charities.

You may ask why the organization of charities appears
in this list. The answer is, because the great national
charity is the Poor-law, and destitution so affects the
public health that the Minister of Health must necessarily
be the Minister of the Poor-laws also. The famine fever
of 1847 terribly exemplified how weighty is the connexion
between wide-spread destitution and national disease. In
that time of panic, error, and confusion, the poorhouses of
Ireland became the great foci of disease, owing to their overcrowding, the evils of which were increased by the congre-
gating of masses of people at public works, from a mistaken
politico-economical theory, and also at depôts for the distribution of food. To these circumstances mainly Dr. Graves
attributes the spread of fever in 1847. In the report of the
Poor Law Commissioners, it appears that the total number
of deaths in the workhouses of Ireland in the week ending
the 14th of April, 1846, amounted to 149. The total number
in the corresponding week in 1847 was 2,706. In April,
1846, the inmates reckoned 50,861. In April, 1847, it was
106,888. The numbers in the union fever hospitals increased
in the one year from 864 to 8,931. In April, 1846, the
weekly rate of mortality in the poorhouses of Ireland was
one in every thousand inmates. Take the city of Cork,
where, as in other places, the horrible system of overcrowding prevailed. From December, 1846, to April, 1847,

2,130 persons died in the buildings comprising the union workhouse. In it the inmates were put three, four, and five in a bed, and in the convalescent ward there were 44 beds for 125 persons.

The deaths in January reckoned . . . 329
,, February 606
,, March 672
,, April 523

Two thousand one hundred and thirty deaths in four months in one over-crowded union workhouse! Similar circumstances occurred in the report of the Census Commissioners for 1851. The deaths in the workhouses, auxiliary workhouses, and workhouse hospitals during the year 1847, as returned by the masters and medical attendants, amounted to 66,890. The general total for the ten years of 1841 and 1851 was 283,765—a number which yet does not represent the total mortality in the workhouses during the ten years, but which gives an annual workhouse mortality during that period of more than 28,000. Can any argument be stronger than this to show the connexion between destitution and disease; any evidence more overwhelming and appalling to prove the want of an enlightened medical police, and to show that the public health must be one of the chief cares of the State? Had there been fundamental laws, as indicated by the author I have quoted, enacted on the true principle of sanitary legislation to help the people to do, not that which they can do, but what they cannot —ascertaining what hindrances there were in the way of the people's health, and removing those which they could not remove for themselves, this country would not have been stained with such fearful results of mal-administration and of ignorance of all the sanitary laws.

Cases of famine-collapse from the country occurred in Dublin. These people had all a strange similitude in feature and expression. Passive and uncomplaining, they seemed to wish only for rest and warmth—they asked for neither food nor drink, but waited for death, silent and cold, and their bodies exhaling an odour as of the tomb. The first impulse was to give them nourishment and stimulants freely, but it was soon found that such a course was rapidly fatal, while all those who were fed like infants for days, and recovered from the collapse, became the subjects of the most rapid and malignant typhus.

But there is another matter relating to this epidemic as showing the necessity of a more severe State control. Emi-

gration then received a new impulse, and the people fled
from their native shores in tens of thousands, as if a curse
had fallen on the land. The overcrowding of the emigrant
ships was frightful, and followed by the inevitable result—
malignant typhus. In 1847 the number of emigrants to
America was double that in 1846, and the ships were not
only crowded, but packed with passengers. In almost all,
as in the workhouses, typhus, engendered by overcrowding,
broke out on the passage. The number of emigrants flying
from disease on the land to meet it on the water was in that
year at the lowest computation 74,539, and the mortality
was greater than on shore. Of these, 5,293 died on the passage, 8,000 reached America in fever ; and if the numbers of
those who died at sea be added to the deaths in the quarantine hospitals, there is a total of 9,786 deaths. Could the
land and the sea give up their dead, what a history would be
revealed of ignorance, mal-administration, suffering, despair,
and wholesale slaughter in the nineteenth century ?

The subject of Preventive as compared with Curative
Medicine is to a great extent a modern one. The latter is
at least as old as the Egyptian dynasty, and the archaic
times of Greece, and has received in modern times a great
development. But as regards its effect on the early mortality of certain epidemics, it can hardly be said that as
yet Curative Medicine has greatly increased the proportion
of recoveries. It is true that in a large majority of cases it
has been used at a disadvantage, or it has not been employed
at all. But we may point to Preventive Medicine as a
development of it. The progress of the physical sciences
in modern times, the introduction of their study into
university education, and their co-option into medical
curricula have influenced the minds of professional men.

Disease has been studied as a part of Natural History, and
we have risen from contemplating its effects to researches
as to its causes. Though medicine does not rank among the
exact sciences, the influences that affect man, in health and
sickness, have been investigated with the accuracy used in
physical studies. All his relations to air, soil, water, food ;
all the conditions which influence him in connexion with
heat, light, moisture, electricity, have been studied and
compared, while the power of statistics has been applied to
the laws which influence his birth, development, inheritance
of disease, strength, and longevity ; also the influence of occupation, the state of civilization, the social and moral
standard of population, and the deathrate in different
countries and places, as compared with that of birthrate.

If a comparison be made as to the relative value of these branches of medicine to the world, I believe it will be seen that Preventive has, or will have, a larger influence for good than Curative Medicine. And you must not forget that the Preventive will act in lessening the necessity of the Curative Medicine.

Preventive Medicine may be looked at from many points of view, but for our present purpose two may suffice—one, the removing the supposed causes of diseases, whether affecting the individual or giving rise to endemics, or epidemics; and the other, those measures which are to promote the physical well-being of the community. A comprehensive sanitary law should embrace both these objects. Here it is right to say that for many years after the commencement of the sanitary movement in England, many of the measures adopted were at the best of very doubtful utility, especially as regards the prevention and extinction of epidemic disease. In fact, many modern sanitarians with but a smattering of knowledge, and with imperfect powers of reasoning, have done much to throw ridicule upon a great subject. But this was to be expected. In every department of human knowledge there have been, and still are, camp followers of science more ready to theorize than to investigate, whose dogmatism is only equalled by their ignorance, and who have adopted some special line of investigation without any previous training or discipline of the mind. These men, in the words of Curran, "hop with airy and fantastic levity over fact and argument, and perch on assertion which they call conclusion." They have too often acted as if the causes of endemic and epidemic diseases were few, easily understood and definitively settled. When cholera first reached London, appearing in a convict ship at Woolwich, it was attributed to a sewer which emptied itself into the Thames just opposite to the stern of the vessel, and immediately that war against sewers, conveniences which unfortunately we cannot do without, received a great impulse, and has continued for nearly half a century. Sewers, streams, rivers, and damp localities, collections of refuse, not alone of putrefying animal or vegetable matters, but of materials in no way offensive have been in turn accused as having been manufactories of disease, not of cholera alone, but in an enterprising mercantile spirit in a great variety, so as to suit the market. Scarlatina, measles, fever, smallpox, have been supposed to be thus generated.

The tendency to attribute complex phenomena to a single cause is rarely seen in right-thinking men. In a

lecture on State Medicine which I last year had the honour
of delivering before the University of Dublin, I quoted these
words of a very learned man. In speaking of epidemics he
says :—" The supposition of a single cause is quite unsup-
ported by nature—every animal, every plant, every rock,
requires for its production the co-operation of many causes,
and probably of many we have not yet discovered. All
nature depends ultimately on a single cause, but it has
pleased that Almighty Cause that the effects which concern
us immediately should arise from the co-operation of several
of His creatures."

Now, to put the matter in a simple way, the favourite
doctrine was, that disease, endemic or epidemic, proceeds
from a cause which is preventable—namely, dirt, foul water,
foul air, foul dwellings, foul habits, and unwholesome food.
And so the question may be asked, what is dirt ? Is it confined to race, latitude, or climate ? Does it always produce
similar effects, or is it capable at one time of causing plague,
at another cholera, or fever, smallpox, or scarlatina ? Is it
always detectable by the eye, or even the nose ? and are
there conditions in which it is nearly, if not altogether innocuous ? Is it an entity, capable of analysis ? are there
conditions in which it is not preventable ? or is its final
cause the furnishing to certain minds a ready explanation of
that which is truly inexplicable ? There is no proof that
dirt, in the common acceptance of the term, ever by itself
gave rise to a single specific disease. If people live in fresh
air, and are not overcrowded, have pure water to drink, and
are not in destitution, dirt does not necessarily make them
sick. Neither, under the same circumstances, are bad smells,
especially if the people are used to them, necessarily ex-
citants of disease. I recommend some of our exclusively
detergent sanitarians to visit the north of Ireland at the
season of flax steeping, when they may inhale a wide-spread
odour which some hold as the *facile princeps* of bad smells.
Yet no consequent annual pestilences ravage that prosperous
country. Our own Liffey, when the weather is warm and
the tide low, is apt to exhale a remarkable perfume, yet Dr.
Grimshaw's fever map of Dublin shows that the quays are
not the habitats of fever. Remember, the inhabitants enjoy
an abundant ventilation, there is no over-crowding, and
little, if any, destitution. I have heard a gentleman who
for many years lived on Upper Ormond-quay, and the oldest
practitioner in Dublin, say that nothing was more remarkable than the longevity of the inhabitants.

It is important to know that there are some antagonistic in-

fluences to the evil effects of dirt, and also that, taken by itself,
it is not the cause or source of epidemic disease. Speaking of
the condition of the farm-houses and steadings of a district in
Ireland, Dr. Pratt observes, that if dirt alone was the common
generator of fever the country would long since have been
desolated from sea to sea. Yet in his district in the north
of Ireland fever is almost unknown. In a town in the south
of Ireland a careful survey gave these results. In one part
of it, measuring twenty-five acres, resided 700 families, including 4,000 persons. Every house had its dungpit,
averaging ten cubic yards in extent, so that in the twenty-
five acres there were at least 7,000 cubic yards of decomposing matter. The lower rooms of some of the houses were
packed with manure, heated and steaming, yet this town
has always been a healthy place, much freer from fever
than any other town in Munster; and the question arises,
can such a condition give different results in different places,
or at different times? Ireland has been from time to time
afflicted with epidemics of fever, but the recurrence of such
is irregular and inconstant, while their supposed causes are
too constant, not alone in towns, but in the isolated dwellings in country districts. Why should these remain long
free from fever while the exciting cause, if it be so, is constant? Or again, why should the character in each epidemic
be peculiar, while the exciting cause remains the same? "I
cannot give my assent," says Dr. Graves, "to the benefits
that are supposed to accrue from opening the sewers and
whitewashing the houses in the poorer parts of cities. It is
true that obstructed sewers give rise to disgusting nuisances,
and soiled exteriors are offensive to the eye. But the causes
of epidemic disease escape the scrutiny of both nostrils and
vision, as is proved by the fact that the worst parts of most
capitals of Europe, however abounding in all sorts of
abominations, do not give rise to either typhus fever, plague,
or cholera. Filth is the outward and visible sign of poverty,
and, like poverty, is itself an evil; it oftener accompanies
than causes disease; otherwise, as I have said, every
capital in Europe would contain within its precincts many
self-supporting manufactories of pestilence. I have always
been of opinion that poverty is more injurious to health
than dirt; that its prevalence entails disease—sporadic
disease — from many obvious causes, and increases its
spread."

In speaking of the attempt to keep the cholera out of
Dublin by enforced cleanliness, for which the Sanitary
Committee obtained power to levy a cess, Dr. Graves says,

referring to a certain parish—"If such tax be levied only off
the rich, it will amount to almost nothing; if it is to be
wrung from the limited resources of the thousand pauper
roomkeepers who inhabit that quarter of the city, then,
indeed, will they have reason to curse the day that the
means for purchasing food were diminished by contributions
intended to remove nuisances, and in vain would they be told
that they must pinch their stomachs for the general purposes
of cleanliness. The Sanitary Committee may, by the force
of perseverance and of the law, effect many of the objects
they have in view; but I fear, even if they entirely succeed,
they will have but converted the wretched dwellings of the
miserable inhabitants into whited sepulchres—the abodes
of contagious maladies, on account of the entassement or
crowded state in which the poor necessarily live. If humanity
strives, therefore, in its visits to the haunts of misery to
prevent the spread of contagion, it must pluck the inmates
from within those bounds, distribute them over a large space,
where the same number that now inhabit rooms may occupy
large houses, and may have the use of nutritious food; but,
alas! this, the only true method of relief, will require something more expensive than the broom and the brush, and
those who are so loud in recommending open sewers and
whitewashing as sovereign prophylactics will perhaps shrink
from contributing their share of that poor-rate or money for
relief, which alone can snatch the pauper population from
the hands of the destroyer."

These weighty words must not be forgotten. The real
antagonistics to any successful Preventive Medicine are
poverty and destitution, with their long train of evils—
ignorance, apathy, insufficient and improper food, filthy
habits, overcrowding, bad ventilation, insufficient clothing,
the living in ruined and neglected tenements, the destruction
of proper pride and the blessed influence of home. This is
the history of cities in which the wealthier inhabitants
gradually forsake the older quarters of the town, which, in
their turn, become ruinous, and of course the homes of misery
and disease.

When we hear a certain class of sanitarians speaking of
Preventive Medicine, some might suppose that the origin of
epidemic diseases, which we have the authority of Humboldt
for saying is one of the most difficult problems in the world,
had been solved. But it is not so, and all measures based
on such an assumption are of more than doubtful application.
The actual origin of specific diseases remains undiscovered.
The origin of the plague, yellow fever, typhus, smallpox,

scarlatina, influenza, is not determined. Nothing is known as to the essence or chemical composition of the virus, nor how one poison differs in offence from another, nor whether the difference is a physical or a vital one, a difference involving qualities which cannot be weighed or measured, or analysed. Little is known as to why it is that one individual will resist contagion and another not. Here the anatomist is at fault. Nothing is known as to why in a single family, exposed to the same exciting causes, there shall be a variety of fever, nor why in the relapse cases the type may be very different from that of the first illness. Of the diseases themselves our knowledge is but a negative one, telling us not what they are, but what they are not.

Fortunately practical sanitation can go on though these obscure questions remain unsettled. Though as yet we know little, if anything, of the essence of epidemic diseases, or why they arise, spread for a season, and then spontaneously subside, we do know that their malignity, their contagiousness, their spreading, and their mortality are increased, sometimes to a frightful extent, by all things that depress the vigour of a population—in a word, by those influences against which sanitary reform has to contend. The list of causes, independent of epidemic disease, which damage the general health of the community, is a long one. The parent of many others is destitution, with its consequences. But to prevent destitution in masses of men, and to promote their prosperity, is the province of the social rather than of the sanitary reformer, who has to deal chiefly with the effects rather than with the causes of destitution, though it is certain that disease and destitution may be and often are reciprocally cause and effect. The origin of specific acute disease being undetermined, it follows that a direct action of Preventive Medicine will be in regard to contagion, by exclusion, if possible, and if we fail in that, by diminishing the number of foci of contagion—that is, by separation of the sick from the healthy. Now, this branch of Preventive Medicine seems applicable to most epidemics, for they all appear to spread by contagion. Observe, I do not advocate any exclusive doctrine as to contagion being the sole cause of the spread of epidemics, but that it is a cause—a very important one—is clearly true. Since I was a student, the admission of contagion as a principal cause of the spreading of epidemics is remarkable, and it now seems more than probable that a vast number of acute essential diseases, that is every acute disease that affects the entire system, may be contagious.

Dr. Graves, in his masterly "History of Cholera," has established its spread in Europe by contagion. Professor Haughton has done the same with respect to its ravages in this country; both authors showing that the disease follows the lines of human intercourse. Of late years even typhoid fever, which was held to be non-contagious, is now looked on in a different light by British authors. The observations of Dr. Flint, of America, upon this point are quite convincing. He gives an example of an outbreak of typhoid in North Boston, where the disease was previously unknown. Its importation was due to a single case ; and such were the facts that we must adopt his conclusion, that if the disease was not transported and diffused by contagion, it is necessary to admit a series of coincidences almost incredible. The circumstances embrace every important condition for a fair experiment to test the contagiousness of a disease. Had they been deliberately selected and arranged for a scientific object, they could hardly have been rendered more complete or judicious.

In estimating the effects of preventive measures with regard to fever, in the generic sense of the word, we are never to forget that, whether we look at the individual case or at the entire epidemic, it is under the law of periodicity. How many great epidemics have sprung up, we know not how, in the history of the world, ran their destroying course, and then, spontaneously went out before sanitary reform was thought of; and therefore, in the present day to attribute their cessation solely to any special proceeding is inconsistent with sound reasoning. It may yet happen—and God forbid that its possibility should be denied—that science will discover the essence of these affections, and with that, the means of preventing and extinguishing them. When that is accomplished, Preventive Medicine will be employed directly, as it is now indirectly.

Till that time comes, however, there is a great work for us to do. We have on the one hand to labour patiently, and in a severely inductive spirit, in the study of the natural history of epidemics, and their comparison in various portions of the world, avoiding conjecture, and honestly accumulating and analysing facts, which thus treated will in time crystallize into discovery.

The term "preventable disease" has been long in the mouths of a certain class of sanitarians. I do not object to these sanitarians. I wish we had more of them among us, and I wish that every citizen and magistrate, from the Lord Mayor down, was a sanitarian. Prevention of disease is to be looked on as involving questions of the causes of specific malady, and also of the modes of mitigating its effects. If, on the one hand, a city is threatened with a contagious epidemic, the example of Copenhagen in 1866 shows that it may be virtually saved from its entrance, while, on the other hand, Preventive Medicine may put a stop to epidemic disease.

In his paper on Comparative National Health, Dr. Acland, speaking of the Indian cholera, says :—" A true and scientific man is always hopeful, is ever ready to pursue a work to the end of his life without attained rewards, and to help younger men in collecting facts and drawing safe and sound conclusions, though he himself may never live to know them." We have, on the other hand, dealing with that which is ascertained, to labour for the physical good of the community, and to strive to keep it in the best condition, so as not only to ensure its health, happiness, and prosperity, but also its power to resist disease, whether it be the pestilence that walketh by night, the epidemic disease which is generally preventable, or the effects of chronic or hereditary ills.

Let me not be misunderstood as slighting or decrying public cleanliness, the importance of which to public health requires no new advocacy. I only seek to show that the want of it is but one out of many causes of general injury to the well-being of the community. The existing state of our city, at least in its older and more decayed portions, is simply deplorable. The inhabitants are too ignorant to abate nuisances, and too poor to get rid of them, though they suffer from them. They are helpless, and exemplify that principle of fundamental law that sanitary legislation should aim at helping, not those who can help themselves but those who cannot.

That the vital condition of the population of Dublin is greatly below par I can state safely on the authority of the Registrar-General for Ireland.* There is no surer test, perhaps, of the prosperity and physical health of a population than the height of the birth-rate in that population. And yet what do we find in the case of Dublin ? That its birth-rate in the last eight years (I have left the year 1864 out of consideration, for it was the first year of registration in Ireland, and the returns are, therefore, in all probability below the truth)—its birth-rate, I say, exceeds its death-rate by only one per thousand of the population annually. The excess of the birth-rate over the death-rate during the same eight years has been in Edinburgh 8¾; in Glasgow 10¼ ; and in London 11¼ per thousand of the population annually, against 1 in Dublin. For every 100 persons who died in

Edinburgh during those eight years, 132 children were born ; in Glasgow, 133 ; and in London, 147. But in Dublin for every 100 deaths there were only 104 births. Again, taking into consideration the relative densities of the population— an all-important factor—we find the death-rate of Dublin to contrast unfavourably with that of the cities I have mentioned. It was during the eight years 1865-72, on the average 26·28 per thousand of the population annually, while in Edinburgh it was 27·25 ; in Glasgow, 30·88; and in London only 24. Now, at the middle of the period of eight years I have chosen for analysis, there lived on every acre in Edinburgh 40 persons ; in Glasgow, 89 persons ; in London, 40 persons ; and in Dublin only 33 persons. For every 100 persons who lived on an acre in Dublin, 121 persons lived on an acre in London and Edinburgh, and no less than 273 in Glasgow. With this sparseness of population, the death-rate of Dublin should be much less than that in London and Edinburgh, and very much less than that in Glasgow, all other things being equal. In actual fact, however, London has a considerably lower death-rate than Dublin. Edinburgh, which is admittedly most unhealthy, but little exceeds the death-rate of Dublin, and Glasgow has a death-rate of only 4 per thousand above that of Dublin.

It is estimated that there are in Dublin *not less than a thousand houses unfit for human beings to live in.* I believe that this estimate is far below the mark. The reports of our nuisance inspectors remind me of early days spent in visiting the poor in the Liberties of Dublin, since which time decay and destitution have been doing their work fourfold in all the poorer parts of our city. The marks of physical degradation in the inhabitants are sickening to look at. The ill-developed frame, the pallid and hollow cheek, the sunken eye, all tell of a population through which endemic and epidemic disease run riot.

Destitution commonly implies over-crowding, filthy habits, bad ventilation, and impure water ; and, putting aside the question of the generation of specific disease, it entails many evils affecting the vigour of the body, its development, and its power of resisting acute and chronic disease. It affects the mortality of children to an extraordinary degree with the influence of hereditary taint, and depresses the entire moral and physical condition of the people. I have dwelt on these topics thus far, although they involve abstract questions ; but, to come to practical matters, there is one great exciting cause among many of injury to public health. I allude to that constant result of poverty—overcrowding.

The poison thus produced is energetic in proportion to the density of the population. "Are we then," asks Dr. Rumsey, "to stop at the palliative stage of sanitary progress ? Can no advance be made in the absolutely preventive direction ; can nothing more be done than is now doing, so imperfectly, to remove the obviously prime cause of the evil—the heaping together vast crowds of human beings in confined areas ?"

It has been shown by Dr. Farre that the mortality of town districts is much greater than that of country districts. In Liverpool, in 1851, the population was 116 persons to one acre, giving 41·7 square yards to each person, but certain wards were very differently circumstanced. In one there were but nine square yards for each person, and in one street, with fourteen hundred inhabitants, the area was but four square yards for each. Typhus fever attacked the inhabitants constantly, and in the worst localities the number of cases amounted to ten per cent. The mortality was in general from thirty-three to forty-two per thousand, while the death-rate in other wards was but from twenty-four to thirty-two per thousand. In five wards in Nottingham the average space for each inhabitant was from eight and a half to eleven and a half square yards, and not only was the death-rate precisely in the order of the density of the population, but the mean age at death decreased from twenty-three to eighteen years in the same order, illustrating, as Dr. Rumsey remarks, the effect of density upon the vital force of the population, with which, as I have striven to show, sanitary reform has most to do.

It would be of great importance if the sanitary inspection of infected houses should determine even by approximation the number of inmates in such dwellings. In Dr. Whitelaw's survey a house containing eight rooms which was situated in Ardee-street, was reported as containing ninety-five inhabitants, that is nearly twelve for each room ; and I know a gentleman who visited that very house in 1870, and found sixteen human beings living in one wretched room.

There is little, if any, evidence to show that the so-called specific diseases are generated by other causes than contagion. This applies to measles, scarlatina, diphtheria, smallpox, and I believe cholera. Of the exciting causes of typhus and of typhoid, something more is known, though of the actual nature of the virus in these diseases we are in utter ignorance. We know, however, that in common with other essential diseases their mortality varies with the epidemic character and the previous physical condition of the sufferer. They are correlative, and under certain states

c

of the system, very probably even convertible. Therefore,
in regard to all these affections the term preventable, except
by measures of exclusion and isolation, is applicable only
in the indirect sense of the word. Preventive Medicine
must be content to deal more with their effects than with
their causes. That typhus is a common result of overcrowding seems sufficiently plain, though the rationale
is unknown; with regard to typhoid, whose correlation to
typhus is so close, there is evidence to show that it has
been excited by preventable causes. But that both these
scourges originate from complex causes at different times and
in different places may be admitted. Not only do the
exciting causes of one seem capable of exciting the other,
but the contagion of typhus may cause typhoid and *vice
versâ*. These are unfashionable doctrines, but I believe
them to be true. The extent of their application is far
greater than what appears at first sight, and as regards
sanitary science, their importance is great, indeed.

I have shown that we know little of the actual cause or
essence of specific disease. Still, measures to hinder the
income of a contagious epidemic may be successful, and
so be classed among those of direct Preventive Medicine.
The efficacy of seclusion of families and communities
during the prevalence of the plague has long been
admitted in the towns of the Levant, where the established
practice is for the Europeans to shut up their houses on the
first appearance of the plague. So uniformly successful is
this practice, that Dr. Russell asserts a case could hardly be
specified in which a secluded family had been affected
without the mischief being traced to some violation of the
rule of confinement. In the severe epidemic of plague at
Marseilles in the last century, the monks, who shut themselves up in their monasteries, escaped. So also Dr. Murtens
preserved the Foundling Hospital at Moscow, which contained a thousand persons, so completely that not one died
of the plague, though great numbers died daily in the city
for some months.

It is plain that in Dublin, as in all maritime cities,
measures should be taken to prevent the importation of
epidemic disease, such as cholera, smallpox, yellow fever,
and so on. In the epidemic of 1828 and 1830 many cases
of yellow fever occurred in Dublin, and since then it has
been directly imported from the West Indies into Sunder-
land. The difficulties of the question must be faced, and
if a strict quarantine cannot be established between two
countries so close as Great Britain and Ireland, and with

such a constant passenger traffic, let us do the next best thing. The efficacy of quarantine has been doubted, because contagious diseases have their latent period in the system. Disease may be actually working in a person coming from an infected district who may appear in good health and sicken after landing. That this is a difficulty is not to be denied, but it furnishes no argument against making the effort to prevent the spread of imported disease.

Let us put aside these cases of possible latent disease, and deal with those in which it is perfectly formed and unmistakable. Dr. Burke, in his paper read before the College of Physicians, animadverting on the public apathy shown as to the importation of smallpox, as contrasted with the feeling which existed respecting the importation of the cattle disease, says it affords another illustration of the principle that man loves gold more than life. It is plain that some preventive system should be resorted to where manifest disease has shown itself aboard ship. Were there no other reason for revision of the sanitary laws, the fact that such a system of prevention has never been brought to perfection in this country would be sufficient.

In the report of the Poor Law Commissioners for this year is given the correspondence with the authorities in Derry in reference to a case in point. A vessel was in harbour with actual smallpox on board, but the powers of the authorities seemed very imperfectly defined, or at least understood. The Poor Law Inspector and the Medical Officer proceeded on board to explain to the captain the arrangements made at the Union Fever Hospital for the reception of the men; but the captain declined their assistance, the officer of health of the city being in charge of the patients. He subsequently changed his mind, and applied to the coast-guard officer for permission to land the men in smallpox, but it then appeared that the ship had been placed in strict quarantine by the commander of the coast-guard, and the request was peremptorily refused. One patient died. The Commissioners endeavoured to find out by what authority the barque, curiously enough named the *Unanima*, had been put into quarantine, and were referred to the Admiralty instructions of the coast-guard, in which instructions it turned out that the officer of the coast-guard had no authority to prevent the landing of any person in smallpox. The Commissioners communicated with Government, suggesting that it was a case for a coroner's inquest, when it was decided that smallpox was not one of the diseases which rendered the ship liable to quarantine. The commander of the coast-guard was declared to have acted under a misapprehension of his duty, the Admiralty regulations as to quarantine not applying to the disease in question, but to cholera. There is a letter from the Mayor of Derry, as to another case of smallpox on board ship, in which he complains that after the patient had been removed to hospital the master of the ship refused to permit the purification and disinfection of the ship. The magistrates and the corporation could not find that there was any authority to punish the master.

All this shows in what a state of confusion the question is. Here was an important town concerned as to the importation of a highly contagious disease, yet there was little unity of action between the various authorities ; and, in the second case, the mayor, the corporation, and the magistrates of the city, are set at defiance by the captain of a coasting vessel.

The whole subject of quarantine and of the separation of the sick in communicable disease, relates to the great principle that sanitary legislation is to be mainly directed towards helping the community to do that which it cannot otherwise effect. Therefore, looking at its object, the public health, it should prevent any individual from interfering, though presumably for his advantage or profit, with the well-being or the health of his neighbours. As a nation, we are jealous of our personal freedom, but the good of the State requires that private interests must be secondary to the public weal.

I believe that in many cases the advance of knowledge will show that repressive laws for the good of the community will be found to harmonize with private interests. The pollution of rivers which carry away the refuse of manufactories, such as calico printing and paper works, exemplifies not alone the necessity of State control over the nuisance, but the advantage of repressive legislation to the manufacturer himself. Some short time since an influential deputation of calico printers waited on the Minister to show that great injury would be inflicted upon trade if his proposed measures in relation to the pollution of rivers were to become law. After stating their case the deputation, with the exception of one member, retired, who then said, "Though I felt compelled to come with this deputation I wish to say that you are to place no value on their representations. I have myself for some time done all that the proposed law requires. We save all the refuse, and return the water to the river in a state of greater purity than when it was first used. Our saving thereby amounts to thousands a year." In the alkali manufacture the pollution of the air by muriatic acid, and in the iron and other works by smoke, are analogous cases.

All that is wanting both for public health and the benefit of the manufacturer, is that the Government, being guided by scientific advice, the most eminent advice—as was had in these cases, should enact and firmly administer, laws for the advantage of the community. In the iron works alone, the waste of fuel has been in some cases calculated at eighty per cent., all of which goes to the deterioration of the public health. In speaking of contagion as connected with Preventive Medicine, the subject of disinfection, which will be fully dealt with by Dr. MacDonnell, must be noticed. If it be asked how far chemical, &c., agents destroy or modify the contagious principle, the answer with reference to sanitation is unsatisfactory and uncertain. Dr. Cameron has ably shown that the employment of these measures as they are commonly used is not to be trusted. He shows that chemical agents will destroy the life of the infusoria, but to produce this effect they must be used in a far more concentrated form than that too often employed by sanitarians. If contagion depends on germs these may be held to be imbued with that latent, as distinguished from the manifest, life of which Carus speaks—the one lasting for an indefinite time, while the other, once commenced, has its appointed development and termination. This is seen in seeds, in animal ova, and roots. "Thus," says Dr. Graves, "the curious fact has been observed of a bulbous root taken from the hand of an Egyptian mummy having germinated when placed in the soil. How happened it that this bulb remained for several thousand years in contact with the fingers of death, without its own vital principle being either extinguished or called into active operation? What power at once preserved that principle and held it in abeyance? And yet so it was. And age after age passed away without summoning into action that wondrous spell which could thus convert this long-enduring tenant of the tomb into the lily of the field, the Scriptural emblem of beauty, and the honoured type of the glories of vegetable life, beside the purity and brightness of whose hues even the raiment of Solomon appeared dull and faded."

Till the germ theory of contagion is established, till we know more of germ life, and of what preserves, what destroys it, and what replaces the latent by the manifest life, I believe that we must mainly trust for disinfection to cleanliness in the *widest* acceptation of the word.

But I must bring to a conclusion this imperfect sketch of a great subject. Those gentlemen with whom I have the honour to be associated will speak with authority on many questions, a few of which only I have been able to indicate :

for example, the exciting causes, if not the origin, of epidemic disease ; the liability to disease, the laws of contagion, and disinfection ; the relation of meteorology to the subject, sanitary engineering, and sanitary law.

The efforts of the Corporation of Dublin, especially in reference to the district drainage of the city, and to the splendid supply of pure water ; the enlightened spirit in which they have met the Sanitary Association, the action of the heads of the University of Dublin as regards State medicine, the establishment of the Sanitary Association, and the union of the Royal Dublin Society with that body for the purpose of instructing the public mind as to Public Health, are all subjects for the earnest congratulation of every well-wisher of our country—of every one who is devoted to her real interests.

APPENDIX.

TABLE showing the *Births* and *Deaths* per 1,000 of the population living, in Edinburgh, Glasgow, London, and Dublin, in the eight years 1865–72 inclusive, with the number of persons to an acre in 1868 and 1872.

Years.	Edinburgh.		Glasgow.		London.		Dublin.	
	Births.	Deaths.	Births.	Deaths.	Births.	Deaths.	Births.	Deaths.
1865,	36	28	42	33	35	24	28	26
1866,	36	27	42	30	35	26	28	29
1867,	36	27	42	28	36	23	26	27
1868,	38	27	42	31	37	24	28	25
1869,	38	30	40	34	35	25	26	24
1870,	38	26	41	30	35	24	27	25
1871,	34	27	39	33	34	25	29	26
1872,	32	26	41	28	35	21	27	29
Mean of 8 years.	36·00	27·25	41·12	30·88	35·25	24·00	27·38	26·38

Years.	Persons to an Acre.	Persons to an Acre.	Persons to an Acre.	Persons to an Acre.
1868,	40·0	88·9	40·1	32·8
1872,	47·1	98·5	43·0	31·3

Ratios of Deaths to Births.		*Ratios of Births to Deaths.*	
	Per cent.		Per cent.
Edinburgh, . . .	75·7	Edinburgh, . .	132·1
Glasgow, . . .	75·1	Glasgow, . .	133·2
London, . . .	68·1	London, . .	146·9
Dublin, . . .	96·4	Dublin, . .	103·8

LECTURE II.

ON THE DISCRIMINATION OF GOOD WATER AND WHOLESOME FOOD.

By PROFESSOR REYNOLDS.

INTRODUCTION.

The impurities often present in water used for drinking purposes, and the numerous adulterations to which articles of food and drink are liable, have frequently formed subjects for able and interesting discourses, in which not only has the nature of each impurity and adulterant been stated but the injury to the public health supposed to result from its ingestion has been placed in a particularly clear light. It is not now proposed to prove either that the practice of adulteration exists, since this is now unnecessary; or that impurities in water or in food can be connected with disease, because the general subject has been dwelt upon in the introductory discourse we have listened to with such pleasure and advantage, and will no doubt be dealt with in greater detail in the succeeding lectures of this course. We now seek only to convey such information as shall facilitate the distinction of pure from impure water, and of safe from unsafe food. For practical purposes it is only necessary for the consumer to be able to say whether or not his water or food supply can be regarded as wholesome. The *identification* of impurities and adulterants is the work of the chemist and microscopist; but it is often possible by very simple means and without the possession of any special skill to ascertain whether or not a particular article is fit for use in the animal economy. For this purpose it is often sufficient to ascertain whether the substance presents the characters which serve to distinguish it, and these characters can in most cases be easily recognised, though it is seldom possible to name the particular impurity or adulterant which may happen to be present without the possession of that "competent medical, chemical, and microscopical knowledge" required by the Adulteration Act. Information of the general kind referred to should be at the command of every medical man, and even of every head of a household; and our aim now is to bring together as much of this class of knowledge as happens to be available at present.

In carrying out the plan just mentioned it would obviously be inexpedient to do more than name the impurities and adulterants which have hitherto been detected in the several substances, and then to state, as briefly as possible, the distinguishing characters of each genuine article. This necessary course will be pursued in the following sections, in the first of which we shall deal with the water supply, and in the succeeding sections with the characters and impurities of the more important articles of food, excluding most of the so-called condimental foods on one hand and alcoholic liquids on the other.

POTABLE WATER.

Its impurities are mineral (lime and magnesia compounds, iron, sulphuretted hydrogen, &c.), and organic (" sewage contamination," including animal and vegetable nitrogenous matter, either capable of, or in process of decomposition, and the products of such change).

GOOD WATER should be free from colour, unpleasant odour and taste, and should quickly afford a good lather with a small proportion of soap. If half a pint of the water be placed in a perfectly clean, colourless, glass-stoppered bottle, a few grains of the best white lump sugar added, and the bottle freely exposed to the daylight in the window of a warm room, the liquid should not become turbid, even after exposure for a week or ten days. If the water becomes turbid, it is open to grave suspicion of sewage contamination; but if it remains clear, it is almost certainly safe.

We owe to Heisch this simple, valuable, but hitherto strangely neglected test. Frankland has shown that it is extremely delicate, and that the production of turbidity under the circumstances named is due to the minute quantity of phosphoric acid present in sewage.*

The Vartry water, *as delivered from the street mains* in Dublin at present, withstands this test perfectly; but it often becomes very impure when allowed to pass through ill-kept cisterns.

---

* The turbidity is caused by fungoid growths.

TEA.

ADULTERATIONS.—There are two chief classes of teas—the green and black varieties. Under the first head are included the Hysons, Twankay, and Gunpowder; and under black teas, Pekoe, Souchong, Congou, and Bohea. Both classes are subject to many serious adulterations at the hands of the exporters, and again on arrival in Europe. Mixtures of different kinds of tea are legitimately made in the course of trade for the purpose of suiting special tastes; but inferior varieties are often dishonestly mixed with the more costly kinds in order to increase profits. Leaving aside the consideration of "tea mixing," we find that green and black teas have often added the leaves of other plants. Those of the plum, sloe, ash, willow, poplar, hawthorn, beech, plane, orange, elm, horse-chestnut, elder, and oak, have been detected. These leaves are dried and prepared by roasting and "facing" so as to resemble genuine tea very closely. The product is sometimes called "Maloo mixture." Facing is used for the purpose of colouring the leaves and increasing weight. The bodies employed are China clay, gypsum, chalk, French chalk, black lead, Prussian blue, indigo, chromate of lead, carbonate and even arseniate of copper, Venetian red, and fine white sand. The powders are attached to the leaf-surface, by a convenient adhesive material.

Spent (exhausted) tea leaves are often dried, coloured with catechu and an iron salt, then faced, and the product mixed with good tea, "Maloo mixture," or Lie tea. The last named substance is made up of the tea and other leaves, sand, or plaster of Paris, bound together by starch or gum, in order to form granular particles which can be "faced," so as to resemble black or green gunpowder.

GENUINE TEA—When placed in a muslin bag and kneaded in warm water for a few minutes should not give up any powder quickly subsiding when the water is allowed to stand. The moist fragments of leaves when spread out should be compared with the very characteristic figure of the genuine leaf given below. The thick looped veins of the true tea-leaves are easily recognised.

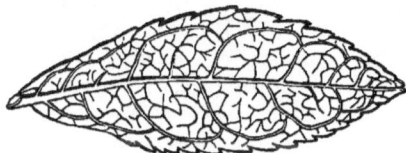

COFFEE.

ADULTERATIONS.—Chicory, acorns, sawdust, roasted roots of various kinds, and grain, tan, croats, lentil seeds, baked livers, Venetian red, burnt sugar. Admixture with chicory is allowable if the compound be truly labelled.

GENUINE COFFEE—Should not cake when pinched between the fingers. If a little be thrown on cold water it floats, and very slightly tinges the water. Adulterated coffee sinks and rapidly colours the water brown.

COCOA.

ADULTERATIONS.—Chicory, cocoa husk, fats, starches, sugar, Venetian red, bole.

GENUINE COCOA.—Should not have a sweet taste, nor red colour. As much cocoa as can be piled up on a threepenny piece, when placed on a square of platinum foil, and strongly heated by a spirit lamp flame should burn almost completely away, leaving a very minute quantity of reddish coloured ash.* The same remarks apply to *chocolate.*

SUGAR.

ADULTERATIONS AND IMPURITIES.—Fine white loaf sugar is rarely adulterated, but coloured sugars sometimes contain chalk, sand, clay, starch, starch sugar, flour, dextrine, plaster of Paris. As impurities, fragments of cane, molasses, vegetable albumen, and sugar mites or *Acari.*

GOOD SUGAR should be free from the least bitter taste, and ought to dissolve completely in water. Loaf sugar should give a perfectly clear and colourless solution; brown sugar, a clear but coloured liquid. If insects are present they float on the syrup, and appear as small specks, which can be easily removed for microscopic examination.

BON-BONS unless when mixed with harmless starch or injurious white or coloured mineral powders produce clear solutions when dissolved in water. If any insoluble residue is left the deposit should be allowed to settle, the liquid poured carefully off, and the powder collected, dried and heated on platinum foil. If white, and wholly combustible, it probably consists of starch. Chromate of lead (yellow), arseniate of copper (green), china clay and gypsum

---

* The microscope is alone able to detect mixtures of many organic bodies, as starches, fats, chicory, &c., in this and other cases. The simple tests given usually serve simply to exclude injurious substances.

(white), and most other injurious mineral pigments give insoluble and fixed powders. Sulphide of mercury or vermillion, though volatile when heated on platinum foil, is easily recognised by affording a heavy red powder on treatment of the sweets with water.

MILK.

ADULTERATIONS.—The chief is undoubtedly water; but skim-milk, annatto, brains, chalk, gum, tragacanth, and other gums; sugar, decoction of white carrots, starch, and turmeric, are stated to be used occasionally.

GOOD MILK should be free from acidity, and when allowed to stand in a vessel ought not to deposit solid matter. When placed in a tall graduated glass cylinder, it should throw up at least 10 per cent. of cream after standing for twelve hours. This is on the whole the least objectionable rough test that can be used.

BUTTER.

ADULTERATIONS.—Water, much salt, starch, flour, dripping, and lard.

GOOD BUTTER should not have a rancid smell. When a quantity is melted and poured into a small narrow phial, and the latter allowed to stand near to a good fire, the milky layer of water that falls to the bottom of the bottle should not form more than one-tenth of the total bulk of fluid. When the melted butter is poured off, the water should not strike a blue colour when shaken with a drop of tincture of iodine.

BREAD.

ADULTERANTS.—Water, rice, potato, and other starches, salt, alum, bone dust, clay, carbonate of magnesium, chalk, gypsum, and sulphate of copper; or impure from bad flour.

GOOD BREAD is sweet and agreeable to the taste. It does not present a mouldy appearance, and ought not to give a thick liquid when steeped in water. If bread becomes soft and sodden on standing it is probably adulterated with rice. When a piece containing much alum is dipped in a very weak solution of the colouring matter of logwood, the bread is quickly dyed of a purple tint.

Good bread ought not to contain more than 38 per cent. of water, and should burn to a very minute ash when heated on platinum foil.

FLOUR (WHEATEN).

ADULTERANTS AND IMPURITIES.—Rice, barley, dari, bean flour, "cones" flour, Indian corn, rye, potatoes, alum, gypsum, clay, ergot, darnel.

GOOD FLOUR should not be acid or musty, but ought to have a pleasant flavour. When a small quantity is burnt on platinum foil (see Cocoa) a scarcely perceptible residue of mineral matter should remain.

As flour containing ergot is poisonous, it is a matter of importance to be able to distinguish this dangerous product of disease in the wheat. We can accomplish this easily by shaking up the suspected flour with a mixture of one part of chloroform and six parts of strong spirit of wine. The ergot, if present in the flour, will float on the liquid and form a brown scum.

ARROWROOT (WEST INDIAN).

ADULTERANTS.—Potato starch, sago meal, rice, gypsum, china clay, chalk.

GENUINE MARANTA ARROWROOT is a dull white powder, which crackles strongly and in a peculiar manner when pressed between the fingers. When mixed with twice its weight of strong hydrochloric acid it yields an opaque jelly. Potato starch, under similar circumstances, affords a transparent jelly. When burnt on platinum foil arrowroot should leave a scarcely perceptible residue if unadulterated with mineral powders. A fragment of iodine placed on a warm plate near to the sample, colours M.A. chocolate brown, sago starch yellowish, wheaten starch violet, and potato starch a dull lilac colour.

MEAT.

BEEF—MUTTON.—Good meat should possess the following easily observed characters:—

1. It ought to be of a full, slightly brownish, red colour; neither of a pale pink tint on the one hand, nor of a deep purple hue on the other. If pink, disease is indicated; and if purple, the animal has probably not been slaughtered, but has died with the blood in it, or has suffered from acute fever.

2. It should have a marbled appearance, from the ramifications of little veins of fat among the muscles.

3. It should be firm, and elastic to the touch, and should scarcely moisten the fingers. Bad meat is usually wet, sod-

den, and flabby, with the fat looking like jelly or wet parchment.

4. It should have little or no odour, and not disagreeable; for diseased meat has a sickly, cadaverous smell. Any disagreeable odour is most easily detected when the meat is chopped up and drenched with warm water.

5. It should not shrink or waste much in cooking.

6. It should not become very soft and wet on standing for a day or so, but should, on the contrary, dry on the surface. (*Letheby.*)

PORK, if unsalted, should present the characters above stated ; but the colour of the meat, if sound, is of very pale red tint. When infested by the dangerous parasite, *Trichina Spiralis,* the meat is usually of a dark colour. Unfortunately, the animal itself can scarcely be detected by the unaided eye ; not so the *cysticerus,* or measle, whose little sac is often as large as a hempseed, and can be easily seen.

SAUSAGES are liable to partial decomposition, and then become poisonous, from whatever kind of meat they may have been prepared. Good sausage meat should be firm, not moist, gelatinous, and vesicular. It should be free from disagreeable smell and taste, and from acidity.

POULTRY.—It is unnecessary to say more under this head than to point out that this class of meat should fulfil the conditions 4, 5, and 6, given above.

FISH should only be used when fresh, and this condition can be easily ascertained. Fresh fish is free from offensive smell, and the flesh is not soft or gelatinous. It may be well to mention that fresh salmon or trout should not only have the well-known pink coloured flesh, but, when the finger is drawn quickly and firmly *across* the fish, the depression so caused ought to fill up quickly, and a corresponding elevation or ridge soon appear.

Sea fish is not tested in this way, but the rigidity of the fish is sufficient to indicate its fresh condition.

The bright red colour of fish gills is a sign of very little importance, as the gills are often artificially tinted.

ISINGLASS.

ADULTERANTS.—Though the best, or Russian isinglass, is an unimportant article of food, it may be well to mention that it is sometimes adulterated with gelatine and with inferior Brazilian isinglass.

GENUINE RUSSIAN ISINGLASS occurs in opaque white filaments, which do not become transparent when placed in water, nor do they swell to a material extent. Gelatine, on the contrary, becomes transparent, and swells considerably. Russian isinglass affords a firm, translucent jelly; the Brazilian variety, for corresponding weights of material and water, does not afford nearly so firm a jelly, and it is much more milky.

VEGETABLES AND FRUIT.

Under this head it is only necessary to say that these articles of food should be invariably used in a fresh and ripe condition.

It is, however, often a matter of importance to be able to distinguish poisonous MUSHROOMS from those that are edible. It may be generally stated that mushrooms which have a disagreeable, styptic taste and a pungent smell should always be rejected. The edible mushroom used in this country has a white top and *pink* gills; as the fungus grows the gills change to a brownish or even nearly black colour. Mushrooms found in open pastures are almost always safe; those found near trees should be avoided.

PRESERVED FRUITS, &c., should not be eaten if mouldy or in a state of decomposition, as evidenced by effervescence or slight frothing, and an unusually acid taste. All preserves, if made in copper vessels, should be tested for copper by stirring a thick bright needle for some time through the preserve, mixed with a little warm water. If, after stirring and standing for an hour or so, the needle, on removal and rinsing with water, is free from any of the well-known reddish deposit of metallic copper, the preserve cannot contain any sensible quantity of the poisonous metal. The same test should always be applied to *Pickles.*

VINEGAR.

ADULTERANTS.—Sulphuric acid, and other mineral acids, water, "grains of paradise," chillies, corrosive sublimate (?). Arsenic and copper as accidental impurity.

UNADULTERATED VINEGAR is allowed by special enactment to contain one-thousandth of oil of vitriol. When paper moistened with vinegar containing this proportion of sulphuric acid is dried before the fire, no charring takes place until the paper is rather strongly heated, but if the proportion of acid is much greater blackening results before the paper seems quite dry. It must be remembered that this is but a very rough and indecisive test. When a piece of clean and bright copper wire is immersed in vinegar, diluted with a little water, and heated nearly to boiling in a glass vessel, the copper quickly loses its colour and assumes a leaden hue if arsenic or mercury is present. Copper may be detected in a fresh sample, much diluted with water, by means of the steel needle, as described under *Preserved fruits.* Pungent substances, " grains of paradise," for example—may be detected by evaporating a quantity of the vinegar nearly to dryness in any convenient porcelain vessel. The residue should not have a fiery taste.

MUSTARD.

ADULTERANTS.—Ordinary " mustard" is rarely free from admixture with one or other of the varieties of flour, turmeric being added to improve the colour. The addition of flour in moderate proportion may be permitted on the score of convenience, but turmeric should not be added. For flour, china clay, plaster of Paris or chalk have been substituted, the colouring material being yellow ochre, or even the poisonous chromate of lead

MUSTARD should not become brown when moistened with a little "spirit of hartshorn," and when burnt on platinum foil, should leave but a small quantity of nearly white ash.

CAYENNE PEPPER.

ADULTERANTS.—Dense flours or starches, mustard, turmeric, ochre, vermillion (?), red lead.

CAYENNE when shaken with cold water, the mixture allowed to stand for a minute, and the liquid poured off should not leave any heavy red powder at the bottom of the vessel. It ought to leave but little ash when burnt on platinum foil.

CHEESE.

ADULTERANTS AND IMPURITIES.—Setting aside such colouring matters as annatto, saffron, &c., we find that the mineral pigments, Venetian red, (red lead ?), are used, and various flours or starches to increase weight.

CHEESE should not be eaten when in a mouldy condition, or when containing "jumpers." It ought not to become blue when touched with dilute tincture of iodine, and it should leave but little ash when burnt on platinum.

LECTURE III.

ON METEOROLOGY IN ITS BEARING ON HEALTH AND DISEASE.

By J. W. MOORE, M.D., DUBL.;

DIPLOMATE IN STATE MEDI INE, AND ASSISTANT PHYSICIAN TO THE FEVER HOSPITAL, CORK-STREET, DUBLIN.

PART I.—*Modern Meteorology.*

BEFORE I proceed to take up the subject of which it is my wish to speak more particularly to-day, it may be well to consider briefly what we mean by the term "Meteorology," in the modern acceptation of the word. And this I deem the more necessary, since many persons are inclined to deny altogether the existence of a weather-science; the triteness of the subject, viewed as a break-ice topic of every-day conversation, having—I suppose—tended to conceal from them the scientific aspect of the study of weather and climate.

Meteorology, in by-gone days, was limited in its application to appearances in the sky, whether atmospherical or astronomical in their character; and this was in strict accordance with the etymology of the word.* But there is no need to remind my audience that the meanings of words change in the lapse of years, and of such a change we have here an instance. The word is now used to denote a branch of natural philosophy which deals with weather and climate; its astronomical connexions are to a great degree severed, while many terrestrial phenomena are included within its vast domain, and are studied and explained under some of its many branches.

If we ask why it is that so important a subject has only of late attracted any general attention, or made any definite progress, the answer becomes easy when we call to mind the complexity of the phenomena with which it deals on the one hand, and the unique character of its relations to science and art on the other. The first point requires no comment. As regards the second, nearly all the sciences and many of the arts are pressed into the service of building a sure foundation for the science of meteorology. Thus mathematics, physics (of which it is itself a branch), chemistry, and biology, all lend their aid to this end, as do also the arts of telegraphy

---

* Τὰ μετίωρα = "Things in the air," "natural phenomena," "the heavenly bodies"—Cicero's "Supera atque cœlestia."—(*Liddell and Scott.*)

and of photography. Until then all these had reached a certain degree of perfection, but little progress was to be anticipated for a science so dependent upon them.

Among meteorologists of old times, the names of Aristotle, Theophrastus, and Aratus, in Greece—of Lucretius, Virgil, Pliny, and Cicero, in Italy—stand prominently forth. But after these worthies had passed away, the sky again darkened for many centuries, until a fitful gleam of sunlight streamed in 200 years ago in the discovery of the barometer by Torricelli. Then came Fahrenheit, Reaumur, and Celsius, the fathers of thermometry; and about a century later, Dalton, Wells, and Daniell, whose names are inseparably bound up with hygrometry. One step further deserves notice:—the investigation of isothermal lines was commenced by Humboldt, and almost perfected by Dové; the latter of whom in his great work on the "Law of Storms," also drew attention to the theory of the winds, a subject which had attracted but little notice from the time when George Hadley, in 1735, published his treatise on the "Trade-Winds" in the *Philosophical Transactions* of the Royal Society.*

So far, the work had been done by individuals; but some twenty years ago, a completely novel epoch in the history of meteorology commenced in the founding of Meteorological Societies in America, and in several countries of Europe. The result was the collection of a vast amount of *trustworthy* material, which leading individual meteorologists such as Dové, in Germany ; Buys Ballot, in Holland; Maury, in America ; and Lloyd, in our own country, were not slow to utilize.

Hitherto very little was known as to the dependence of the direction and force of the wind on barometrical and thermometrical conditions, at least outside the tropics. But in 1854 a countryman of our own, Dr. Lloyd, the present esteemed and distinguished Provost of Trinity College, demonstrated the cyclonic character of most of the gales experienced in Ireland,† and so foreshadowed what is now universally known as *Buys Ballot's Law*—a law on which the whole of modern meteorology turns. As applicable to the Northern Hemisphere, it may be concisely stated as follows :—

" If at the same moment of time there be a difference between the barometrical readings at any two stations within a reasonable

* "On the Cause of the General Trade-Winds." By Geo. Hadley, esq., F.R.S., No. 437, p. 58.
† "Notes on the Meteorology of Ireland." *Royal Irish Academy Transactions,* vol. xxii., Science ; 1854.

distance from each other, a wind will blow on that day in the neighbourhood of the line joining those stations, which will be inclined to that line at an angle of nearly 90°, and will have the station where the reading is lowest on its left-hand side."

In more homely language—"If on any day a person stands with his back to the wind, the reading of the barometer will be lower at all stations on his left-hand than it is where he is at the time."

Thus the *direction* of the wind is determined by differences in atmospherical pressure which are marked by differences in the height of the barometer. But further, the *force* of the wind also is chiefly regulated by the amount of those differences, or by what are called the "barometrical gradients." "The gradients adopted by the Meteorological Office, London, are expressed in hundredths of an inch of mercury per 50 geographical miles."*

In a paper on "Weather Telegraphy," by the present Director of the Meteorological Office, Mr. Scott, the following passage occurs:—

"The immediate result of the law is to show that whenever barometrical readings are lower over any area than over those adjacent to it, the air will sweep round that area as a centre, and the direction of its motion will be opposite to that of the hands of a watch. Conversely the air will sweep round an area of relatively high barometrical readings in the direction in which the hands of a watch move. The former of these motions is said to be *cyclonic*, the latter *anticyclonic*. These names are derived from the word 'cyclone,' the general name for hurricanes and typhoons, in all which storms the motion of the air takes place around an area of diminished barometrical pressure. . . . The actual movement of the air has no reference, either in direction or velocity, to the absolute readings of the barometer at the point where it is lowest, or to the distance of the particles of air which are in motion from that point, but is related almost entirely to the distribution of pressure in accordance with Buys Ballot's Law. The law gives the direction of motion, and its truth for these islands and the adjacent parts of the earth's surface is incontestable."

Having now shortly reviewed the history of meteorology from the earliest times to the present, I will give in as few words as possible an explanation of the climate of the British Isles (1) in summer, and (2) in winter.

We may lay down as aphorisms—

First. That hot air is lighter than cold air.

Secondly. That the rapidity with which the processes of

heating and cooling of air go on is in direct proportion to the amount of aqueous vapour contained in that air—dry air becoming heated or cooled more rapidly and more completely than moist air, other conditions being alike.

Thirdly. That, consequently, the air over large areas of land, being drier, becomes more rapidly heated in summer, and more rapidly cooled in winter, than air which is in contact with extensive water-surfaces; and,

Fourthly. That the radiation-heating power of dry land is greater than that of water, as also the radiation-cooling power of dry land is greater than that of water.

This group of facts is of paramount importance in climatology.

Now let us apply these facts. Over the centre of the great continent of Europe and Asia the air in summer will become much warmer than that over the Atlantic Ocean to the west, and over the Pacific Ocean to the east; the barometer will consequently fall over Russia, Siberia, and other inland countries, the isobars, or lines of equal barometrical pressure, curving round the point of lowest pressure, while it will remain tolerably high over the oceans I have mentioned. In accordance with Buys Ballot's Law a circulation of air will commence round the barometrical depression thus formed, a vast *cyclone* will become developed, the winds blowing against the hands of a watch, from S.W. in India and China (the S.W. monsoon), from S., S.E., and E. in Japan, and North-eastern Siberia, from N.E. and N. in North-western Siberia, from N.W. and W. over most of Southern Europe and South-western Asia.

In winter, on the contrary, the air over the central districts of Europe and Asia, rendered dry by the intense heat of summer, and its accompanying excess of evaporation, will become rapidly chilled to an extent of which we can form scarcely any idea, the air will be condensed, and the barometer will rise, while pressure will diminish over the Atlantic and Pacific Oceans, where the temperature is perhaps 60° or 80° higher. Thus, conditions, just the reverse of those observed in summer, will be established—an immense anticyclone will be formed, the winds circulating round and *out from* the centre of high pressure in a direction with the hands of a watch, blowing from N.W. and N. in Japan and China, from N.E. in India (the N.E. monsoon), from E. and S.E. in Russia and Southern Europe, from S.W. in the British Isles, and from W. in Northern Russia and Siberia.

I have said that we can form but little idea of the enormous changes of temperature which take place in Central and Northern Asia between the seasons of summer and winter. But that these changes are sufficient to produce the great variation in barometrical pressure on which depends the varying wind-system of the continents of Europe and Asia in those seasons may be easily shown by a comparison of the range of temperature between July and January in an insular climate like our own, and at Yakutsk in Siberia, which is situated close to the centre of lowest and highest barometrical pressures in those months respectively. At Dublin the mean temperature of July is about 60° F., of January about 40° F.—a range of only 20°. The corresponding mean temperatures at Yakutsk are 74° F. and − 40° F. respectively—a range of 114°. For weeks in summer the thermometer ranges between 80° and 90° at this place, while in winter it may descend 90° below the freezing point of water. Well does Humboldt observe:*—

"The inhabitants of the countries where such *continental climates* prevail seem doomed, like the unfortunates in Dante's 'Purgatory'—

'A soffrir tormenti caldi e geli.'"

Or, as Milton has so well expressed it—

"From beds of raging fire to starve in ice."

In the winter season the predominant winds over Scandinavia are south-easterly, but this apparent anomaly is in fact a beautiful fulfilment of the very laws it seems to contradict. I have said that in winter a barometrical depression exists over the North Atlantic Ocean. It is this which draws the wind from S.E. over Sweden and Norway, in strict agreement with Buys Ballot's law.

It will easily be seen how the summer continental depression influences the climate of the British Isles. Air is drawn from W. and N.W. over these countries, and as this air blows over the surface of a wide ocean, and from high latitudes, it is cool and moist. Do not these two words describe our summer? If all I have stated be true, these ocean winds will prevail chiefly on the W. and N.W. shores of Ireland and Scotland, which will thus have the rainiest and the coolest summer, while this season will be warmer and drier as we go eastward and southward, to the south-eastern counties of England. This is well illustrated in Mr.

* "Kosmos," vol. i., p. 352.

Buchan's Chart* of the Isothermals of the British Isles in July.

It is not necessary to consider at length the influence of the winter-system of barometrical pressure on our climate. During the earlier winter months a great stream of warm, very moist air, as a rule, flows north-eastward and northward over these islands round the Atlantic depression, the centre of which lies near Iceland. But this stream does not flow evenly. Along its eastern edge it is in continual conflict with the cold anticyclonic air, which is travelling westward from Russia and Siberia, and immense volumes of the latter are ever and anon rushing in to supply the place of those volumes of the warm air which, owing to their low density, have presumably risen from the earth's surface towards the higher strata of the atmosphere. This conflict between two such opposite currents of air causes our storms, and those violent and rapid alternations of temperature, which, as I hope to show you, are so prejudicial to health in the winter months.

The reason for the occurrence of these alternations of temperature will be explained when we remember that most of these gales, or *bourrasques* as they have been termed, are cyclonic in character, and that they generally cross the British Isles from S.W. to N.E., less frequently from W. to E., and still less frequently from N.W. to S.E. The southerly winds then which blow over the country in front of the centre of the storms are warm and moist, while the northerly winds, which prevail over those districts already reached and passed by the centre, are cold, and after a time dry. No better examples of this can be given than the remarkable gales of December 8th and 9th, 1872, and of February 2nd of this year. In front of the former temperature rose generally to about 50° over the south of Ireland, most part of England, and all of France; while it fell almost to the freezing point over those districts a few hours later when the centre had passed. The second gale I have referred to was accompanied by a range of 18° (Fahrenheit) over the whole of France.

The effect of the warm Atlantic air-current on the Isothermals of the British Isles is well represented in Mr. Buchan's Chart for January.*

Anticyclonic wind-systems sometimes prevail over western Europe, but much less frequently than cyclonic systems. They cause dry, often cold weather, and are much more persistent than cyclones.

* "The Temperature of the British Islands." By Alexander Buchan: *Journal of the Scottish Meteorological Society,* October, 1870.

This notice of modern meteorology would not be complete without some words on the telegraphic system of observation and storm-warnings. In a paper published a few months ago I wrote as follows :—*

"Dové's investigations on the 'Law of Storms,' followed by the enunciation by Professor Buys Ballot of his laws respecting cyclonic wind-systems, were supplemented in the year 1860 by the introduction of telegraphic meteorology, a step in advance, for which, as regards Great Britain, we are indebted to the late Admiral Fitzroy. Of late years this important branch of the science has been further developed and brought to a certain degree of perfection by the extension of the area of observation, and by the co-operation of public departments in most European countries with the Meteorological Office, London. To the agency of the last-named, which has been for some years connected with the Royal Society, and over which a countryman and fellow-citizen of our own, Mr. Robert Henry Scott, Fellow of the Royal Society, has ably presided in the capacity of Director since February 7th, 1867, we owe much of our present knowledge of the wind-systems of Western Europe. Telegrams are now sent daily to the office from three stations in Norway, one in Denmark, one in Germany, one in Holland, one in Belgium, nine in France, one in Spain, one in the Shetland Isles, one in the Hebrides, six in Scotland, five in Ireland, and sixteen (including Scilly) in England and Wales. These telegrams reach the office generally about 11 A.M. The observations are then reduced and discussed, and from them a daily weather report is drawn up, lithographed, and sent out early in the afternoon to many of the London papers. Daily weather charts are also published on the same sheet, and on them are drawn the isobars, or lines of equal barometrical pressure ; the isotherms, or lines of equal temperature ; curves illustrating the general direction of the wind ; notes of the prevailing weather, rain, storm, &c., at 8 A.M., over most of Western Europe."

PART II.—*Influence of Meteorological Conditions on Health and Disease.*

Observations as to the influence of weather upon health are as old as meteorology itself—nay older, if we regard Aristotle as the founder of the science ; for more than 400 years before the birth of Christ, Hippocrates of Cos, the "Father of Medicine," had penned his immortal "Aphorisms," and had written " Πέρι ἀέρων, ὑδάτων, τόπων " ("On Air, Water, and Places,") and " Πέρι διαιτῆς " (" On Regimen.") In these works we meet with passages as applicable to-day as they were twenty-two centuries ago. Let me quote but two or three of these :—

(a) " The changes of the seasons are a fertile source of maladies,

* *Irish Farmers' Gazette,* July 13th, 1872.

and in the seasons themselves great variations of cold and heat, and other things proportionally."

"(β) Some constitutions fare well or ill in summer, others in winter."

(γ) "Different diseases prevail at different seasons, or again subside."

Now mark the hurtful character of sudden changes of the weather :—

(δ) "In any season of the year, should heat prevail at one time, and cold at another of the same day, we may anticipate autumnal maladies."

Again, while writing on *Regimen*, he says :—

(ε) "These persons (of men) are more liable to sickliness in winter than in summer, and in spring than in autumn."*

There is reason to believe that the suggestions thrown out by the Greek physician were allowed to remain almost a dead letter. Certain it is that his doctrines as to the close relation of climatology to medicine became dimmed by the rust of time, and were neglected or forgotten. That in these countries but little attention was given to the subject is evident from the antiquity and popularity of the proverb— "A green Christmas makes a fat churchyard." Even Sydenham stated that a prevailing epidemic ceased on the approach of winter,† but on the whole his observations as to the dependence of disease on season are accurate and well worth perusal.

The first modern paper on the subject was a communication made to the Royal Society in 1797 by Dr. William Heberden, jun., F.R.S., on the "Influence of Cold on the Health of the Inhabitants of London.‡" The author shows that a difference of above twenty degrees between the mean temperature in London in January, 1795, and that in the same month of 1796 (the former being an excessively cold month, and the latter an equally mild one) caused the deaths in January, 1795, to exceed those in January, 1796, by 1,352.

* (α.) Αἱ μεταβολαὶ τῶν ὡρέων μάλιστα τίκτοισι νοσήματα. καὶ ἐν τῇσιν ὥρῃσιν αἱ μεγάλαι μεταλλαγαὶ ἢ ψύχιος ἢ θαλψιος, καὶ τ' ἄλλα κατὰ λόγον οὕτως.—"Aphorisms," sect. iii.

(β.) Τῶν φύσεων, αἱ μὲν πρὸς θέρος. αἱ δὲ πρὸς χειματα, εὖ ἢ κακῶς πεφύκασι.—Ibid.

(γ.) Τῶν νόσων ἄλλαι πρὸς ἄλλας εὖ ἢ κακῶς πεφύκασι.—Ibid.

(δ.) Ἐν τῇσιν ὥρῃσιν, ὅταν τῆς αὐτῆς ἡμέρης, ὀτὲ μὲν θαλπὸς ὀτὲ δὲ ψῦχος γίνηται, φθινοπωρινὰ τὰ νοσήματα προςδέχεσθαι χρή.—Ibid.

(ε.) Ταῦτα τὰ σώματα ἐν τῷ χειμῶνι, νοσερώτερα ἢ ἐν τῷ θέρει· καὶ ἐν τῷ ἦρι, ἢ ἐν τῷ φθινοπώρῳ.—Περὶ Διαιτῆς, Book i.

† Swan's "Sydenham," 1769; p. 9.
‡ "Philosophical Transactions," Vol. lxxxvi., No. 11.

In my remarks on the present occasion I shall confine myself to the more immediate consideration of only two or three meteorological data in their influence on disease and death among the population of our own city. These data are *mean temperature, rainfall,* and *humidity*—the first the most important determining factor in the inquiry, as it is in truth the resultant of many others. My subject-matter I shall draw chiefly from the weekly, monthly, and annual returns of deaths in the Dublin registration district, published periodically, since the beginning of 1864, by the Registrar-General for Ireland, Mr. Donnelly, c.b. I purpose to deal with the subject under three headings :—

I. The influence of season on *Thoracic* and *Abdominal* affections respectively.

II. The influence of season on the progress of epidemics of recent years, namely, (1) *Cholera,* and (2) *Smallpox.*

III. The influence of season on four principal endemic and epidemic diseases, namely, (1) *Measles,* (2) *Whooping-cough,* (3) *Scarlatina,* and (4) *Fever.*

For the humidity-curves for the years 1864–68 inclusive, and for the rain-fall curve for 1864, I am indebted to the observations taken at the Ordnance Survey Office, Phœnix Park. For the remainder of these curves, and for that of the mean temperature throughout the whole period of nine years I am myself responsible.

It is to be regretted that statistics of sickliness apart from mortality are wanting in this country. How much a system of registration of disease is to be desired has been pointed out by Dr. Stokes in an admirable address delivered in this theatre a fortnight since. Already in the sister country local records of sickness have been compiled, and have been used in investigations similar to those whose results I now lay before you. I need instance only the " Weekly Tables of Disease" which are kept at Manchester, and the careful " Records of Sickness" kept in Islington—which last Dr. Ballard* has analyzed in the most perfect manner, so as to lay the foundation for the more accurate study of climatology in relation to health.

However, as such records have not been kept in Dublin, I must be content to deal with mortality alone, and while doing so I would crave your indulgence should I appear unnecessarily tedious or wearisome.

* Eleventh Report of the Medical Officer of the Privy Council, 1868. No. 3.

DIAGRAM I.

Compiled from the Quarterly Returns of Deaths in the Dublin Registration District for the past eight years, and from the Quarterly Returns of Deaths within the Municipal Boundary of Dublin for the year 1864, illustrative of the effect of certain Meteorological conditions on the Death Curves from Thoracic and Abdominal affections.

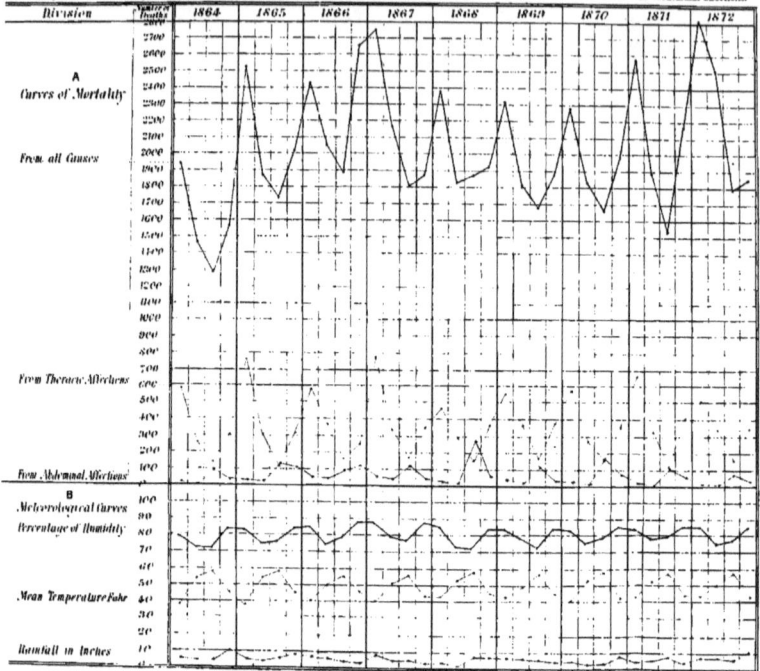

I. Influence of Season on (A.) Thoracic Affections,
and (B.) Abdominal Affections.

At the outset of our inquiry into the influence of meteorological conditions on the death-rate from thoracic and abdominal causes, we may lay down the two following
propositions :—

(A.) *In* Summer *the tendency to sickness and death is chiefly
connected with the digestive organs; diarrhœa and dysentery
being the affections which are especially prevalent and fatal
during this season. In* Winter *a similar tendency is
noticeable in connexion with the organs of respiration;
bronchitis, pneumonia, and pleuritis being the affections
which are principally met with at this season.*

(B.) *In* Summer *a rise of mean temperature above the
average increases the number of cases of, and the mortality
from, abdominal affections. In* Winter *a fall of mean
temperature below the average increases the sickness and
mortality from thoracic affections.*

In proof of the truth of the first of these propositions in
the case of Dublin, I would refer to Diagram I., in Section
A of which curves are drawn, according to the same scale, of
the deaths (α) from all causes, (β) from thoracic diseases,
and (γ) from abdominal diseases, in Dublin during each
quarter since the beginning of 1864. Regarding the first
curve, that of the total deaths, its uniformly greater height
in the *first* quarter of each year at once catches the attention.
Again a depression occurs in the *third* quarter in eight of
the nine years, the exception being 1868, a year of warmth
and drought. We may take it then that the first quarter
of the year is the most deadly, the third quarter the least
so.

Turning to the second curve, that of deaths from thoracic
affections, we find that its summits all occur in the first
quarter, its depressions as invariably in the third quarter of
the year. Another remarkable point deserves notice,
namely, the pronounced influence of this curve in determining the contour of the curve of total mortality, especially in
the winter months.

If we pass now to the third curve, that of the deaths from
abdominal causes, the results are equally uniform. Thus its
summits are reached (with one exception) in the third
quarter, its depressions (without exception) in the second
quarter of the year. The exceptional year was 1866, in the
last quarter of which a fearful epidemic of cholera was
accompanied by a considerable amount of diarrhœa.

So far then a dependence of the thoracic death-curve on season is observed—the season of low mean temperature, of a high per-centage of humidity (saturation being 100), and generally of an increased rainfall, coinciding with the summits of this curve. Similarly, the abdominal death-curve is influenced by a season of another type—a high mean temperature, a low per-centage of humidity, and, generally, a smaller rainfall, coinciding with the summits of this curve.

If the second proposition be true, the death-curves should be intensified by *extreme* mean temperatures—that of thoracic affections in winter, that of abdominal affections in summer.

Now if we look at the thoracic death-curve, we shall see that in the first quarters of 1865, 1867, and 1871, it was very acute indeed. In 1865 the mean temperature of the first quarter was 2·3° below the average (41·6°) temperature of that period during the nine years 1864–72; in 1867 it was 2·2° below the same, and in 1871 it was 1·1° above the same. It appears then that in the two former years a depression of 1° caused the death of about 100 persons from thoracic affections alone. The anomaly observed in 1871 is easily explained by a reference to the mean temperature curve. It will be seen that, although indeed the mean temperature of the first quarter of 1871 was about a degree over the average, yet that of six months ending March 31, 1871, was considerably below the average temperature of this long period. The accumulated effects of continued cold more than compensated for the comparatively mild temperature of January and February, 1871.

The lowest summits of the thoracic death-curve coincide with the mild winters of 1867–68, 1868–69, and 1871–72.

Apart from the cholera year 1866, the abdominal death-curve shows high summits in 1865, 1868, and 1870. The majority of my audience will not have forgotten the wondrous September of 1865, with its July temperature, and absolute drought; the burning suns of the summer of 1868; and the glorious weather of July, August, and September, 1870. The mean temperature of the third quarter of 1865 was 59·8°; of 1868, 59·3°; and of 1870, 59·2°, against an average temperature for the same period of 58·4°. In the summer of 1868, especially, the mortality from diarrhœa assumed alarming proportions, and in the Weekly Return of Births and Deaths in Dublin for August 22nd of that year, the Registrar-General writes :—

"The number of deaths from diarrhœa *registered* during the week amounted to 49, showing an increase of 23 on the number

DIAGRAM II

Showing the Yearly Mortality and Meteorological Conditions, 1864-72, in Dublin.

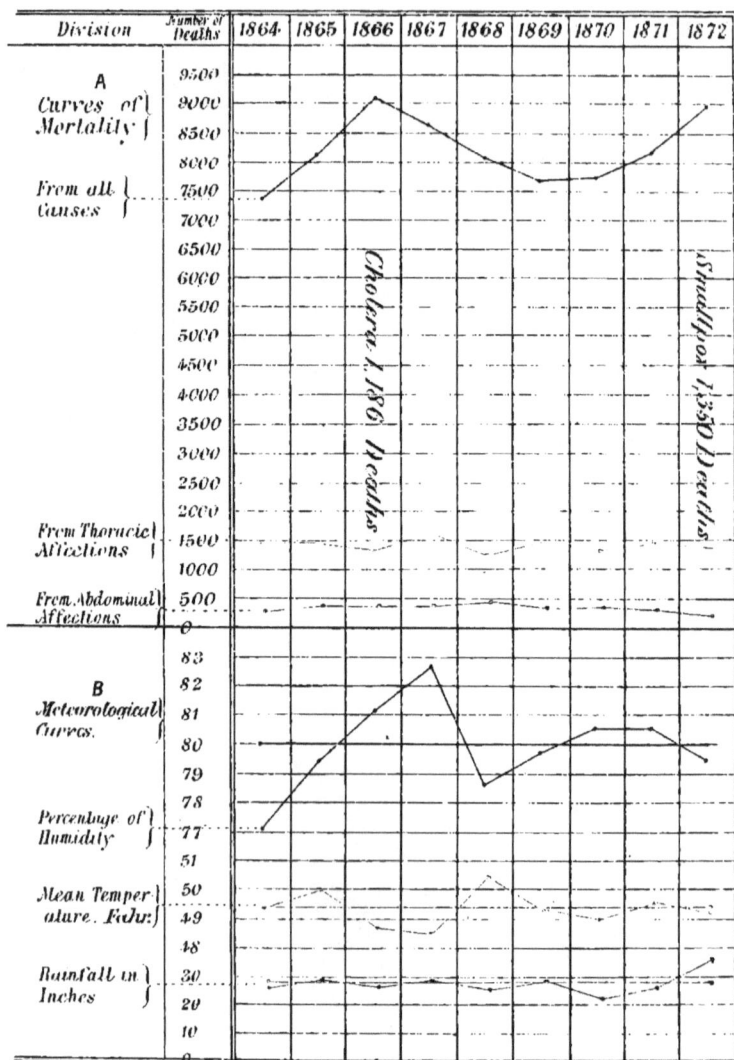

Division	Number of Deaths	1864	1865	1866	1867	1868	1869	1870	1871	1872
A Curves of Mortality.	9500									
	9000									
	8500									
	8000									
From all Causes	7500									
	7000									
	6500									
	6000									
	5500									
	5000									
	4500									
	4000									
	3500									
	3000									
	2500									
	2000									
From Thoracic Affections	1500									
	1000									
From Abdominal Affections	500									
	0									
B Meteorological Curves.	83									
	82									
	81									
	80									
	79									
	78									
Percentage of Humidity	77									
	51									
Mean Temperature. Fahr.	50									
	49									
	48									
Rainfall in Inches	30									
	20									
	10									
	0									

Cholera 1 186[ ] Deaths

Smallpox 1,350 Deaths

registered during the week preceding, and being 35 more than the average deaths from this disease in the corresponding week of the four previous years."

Providentially in the middle of August a copious fall of rain, amounting in one week to 303 tons of water on every acre took place, and within three weeks the plague was stayed!

I have given an example from the Registrar-General's Weekly Return of Births and Deaths of the effect of heat and drought in augmenting the mortality from diarrhœa. Let me quote from the same authority respecting the effects of the cold of January, 1867 :—

" The deaths from bronchitis (*registered* during the fourth week) amounted to 80. The increased mortality from this disease is due to the extreme cold which prevailed; the mean temperature for the first three weeks of this year was 29·8° ; during that period the thermometer fell to 2·8° on the 3rd instant, to 19·2° on the 12th, to 12·5° on the 16th ; whereas during the corresponding three weeks of last year the mean temperature was 43°, and the deaths from bronchitis *registered* during the fourth week were only 31."

In the following week, that ending February 2nd, 1867, the deadly effect of this unusually cold wave was fully felt, no less than 119 deaths being referred to thoracic causes out of a total of 286 deaths registered.

The apparent anomaly of the high thoracic death-curve and of the comparatively mild mean temperature of the first quarter of 1871 has been alluded to and explained. That the explanation given is not unjustified by fact will, I think, be evident from a perusal of the returns of mortality in the weeks of January, 1871. A vigorous frost set in on December 21st, 1870, and from the 24th of the month until January 2nd, following, the ground was covered with snow. In the Registrar-General's return for January 7th, 1871, we read :—

" The very cold weather has increased the mortality from diseases of the respiratory organs. The average number of deaths from bronchitis in the corresponding week of the previous seven years was 29 ; during the past week 53 deaths from bronchitis, and 12 from pneumonia or inflammation of the lungs, were registered."

In Diagram II. I have given, as it were, a *résumé* of the yearly results, but a few words of explanation will be necessary. And in the first place, we must remember that the curve of total mortality is influenced by two factors especially—the presence of an epidemic, and the meteorological conditions. To the first of these my friend, Dr. Grimshaw, will direct attention at an early opportunity, and I therefore purposely avoid dilating upon this subject at length, knowing well in how much worthier hands I leave it. Suffice it to say, that the abnormal curve in 1866 was due to cholera, which caused 1,186 deaths, while the high curve of 1872 was due to smallpox, which proved fatal in 1,350 instances within that year.

But, apart from the effect on the general death-curves of these epidemics, the annual death-curve from thoracic affections appears to be anomalous in 1866, and in 1870, 1871, and 1872, as regards its relation to the mean temperature curve. "*Appears to be,*" but is not. The low mean temperature of 1866 was due to no extreme winter-cold, but to a remarkably cool spring and summer—the mean temperatures of these seasons being 50·9° and 56·3°, against the average temperatures 52·5° and 58·4° respectively; 1870 was mild until the fourth quarter, which was much below the average as regards mean temperature (42·5° against 44·9°), but cold weather tells only after some continuance in early winter, and so we find the increased mortality thrown into the early months of 1871, which were in themselves mild.* As regards 1872, the high mean temperature of the first quarter (44·0°) is quite sufficient to account for the fact that the thoracic death-curve was comparatively low.

The annual death-curve from abdominal diseases presents an anomaly in 1866, 1867, and 1870. The epidemic constitution which accompanied the cholera of 1866 is more than sufficient to account for the comparatively high death-rate from diarrhœa in the two former years, while the hot and dry summer of 1870 more than counterbalanced the low mean temperature of the whole year in question, which was determined by a cold winter quarter and a very cold autumn.

I may be asked, "But is not cold weather 'bracing' and tonic' to the system?" Yes, it is so, doubtless, to the young and strong, to those in robust health and in the prime of life. These classes of the community are invigorated by the cold of winter, and may set the heat of summer at defiance. But far otherwise is it with " the *very*

* An illustration of this may be culled from Dr. Stark's report to the Registrar-General of Scotland for the year 1865 (Eleventh detailed Annual Report of the Registrar-General of Births, Deaths, and Marriages in Scotland, 1868, p. xlv.):— "The reason why March," writes Dr. Stark, "with its higher temperature, had nearly as many deaths as January, was chiefly owing to the fact that the prolongation of the cold weather, even though the cold did not increase in intensity, increased the number of deaths—those whose health was enfeebled by the first accession of cold not being able to withstand the effects of its continuance."

young, the *weakly,* and the *aged."** Children under five
years and the aged go down like grass before the scythe,
when the keen frost-wind or the fiery heat of summer
sweeps across the land.

The appended facts from the Irish Registrar-General's
returns will prove this statement :—

"In 1867, 32·5 per cent. of all the deaths registered in Dublin
were those of children under 5 years of age, and 19·5 per cent.
were those of persons aged 60 and upwards.

"The corresponding per-centages for 1868 were 23·1 and 18·9,
respectively. In the first quarter of 1867—so noted for its intense
cold—the per-centages were : of those under 5, 24·9 ; of those
over 60, 23·5. In the third quarter of 1868—the year of great
heat—the numbers were : of children under 5, 41·5 ; of adults
over 60, 16·9 per cent."

I will close this portion of my subject with the words of
the Registrar-General of England :†—

"When the thermometer falls to the freezing point of water, the
mortality is raised all over the country; and the population of
London is excessively sensitive to cold; thus the corrected average
deaths for the second week of January are 1,550, but the actual
number of registered deaths this year (1864) was 2,427. The
mean temperature of the preceding week, instead of 37·8°, had
fallen to 26·7° ; and the temperature of one chill night (Thursday,
January 7th) had descended to 14·3°, or to 17·7° below the freezing point of Fahrenheit; and 877 lives were extinguished by
'the cold wave of the atmosphere.'"

TABLE I.—Showing the EFFECTS of COLD and HEAT on the
MORTALITY in SCOTLAND during the Year 1868.

SEASON.	Average Mean Tempera- ture of 10 years.	Mean Tempera- ture, 1868.	Difference from Aver- age Mean Tempera- ture.	Death-rate per cent.		Lives saved.	Lives lost.
				Average.	1868.		
First Quarter, .	37·9	40·6	+2·7	2·49	2·26	1,833	–
Second Quarter,.	49·7	51·0	+1·3	2·22	2·12	796	–
Third Quarter, .	55·3	57·4	+2·1	1·90	2·09	–	1,514
Fourth Quarter,.	41·9	41·5	—0·4	2·14	2·22	–	673

* Registrar-General of England in the *Times* of Friday, May 7th, 1869.
† "Twenty-seventh Annual Report of Births, Deaths, and Marriages in England
for 1864," p. xlvi.

II. Influence of Season on (A.) Cholera and (B.) Smallpox.

A. In an admirable report by Professor F. C. Faye, of Christiania,* we read as follows :—

"Most epidemics of cholera, especially those of extreme violence, have occurred in summer and autumn, so that in large towns, where the observations are less influenced by circumscribed local conditions, a direct relation has been established between the rise of the epidemic and that of the temperature. That warm air should be more favourable to an epidemic is explained by the fact that all exhalations, both from the ground itself, especially where it is swampy, and from all vegetable and animal organic matter scattered over its surface, are more rapidly developed ; and as water also acts as a solvent on these materials, the further influence of moisture is easily understood. It has been matter for surprise that an incessant rain has a more healthy influence, although during it the moisture on and near the surface of the ground is very considerable ; but as to this, it has been rightly observed that the exhalations from a purer water cannot be supposed to be particularly hurtful, and that therefore the quantity of water which turns swamps into seas, and in other places washes away impurities, acts advantageously in proportion."

The author adds :—

"The months of August, September, and July, have on the whole been those of greatest mortality, and when the disease has commenced at rather an advanced period in the autumn, it has generally happened that the number of cases has been greater in the milder season than during a falling temperature."

These remarks of the Norwegian Professor are so true and so universally applicable that without further preface I would direct attention to Diagram III., in which I have drawn curves of the deaths from cholera in Dublin by weeks, and of the prominent meteorological conditions of the epidemic period in 1866. As this awful and mysterious disease is, unhappily, so rapidly fatal, we may safely compare the deaths in any week with the meteorological conditions of the week immediately preceding. I may state that on the whole, a comparatively low temperature prevailed throughout the epidemic period, and we may suppose that had this not been so, a far greater development of cholera might have occurred, terrible as the death-rate really was.

The first great mortality (41 deaths) was registered in the week ending September 1st, a week after the mean temperature had attained the height of 61°, and closely following

* "Om Cholera-Epidemien i Norge i Aaret 1853" ("On the Cholera-Epidemic in Norway in the year 1853").

DIAGRAM III

Containing the Death Curve from Cholera in the Dublin Registration District in 1866-67. with the chief Meteorological Conditions of the Epidemic period.

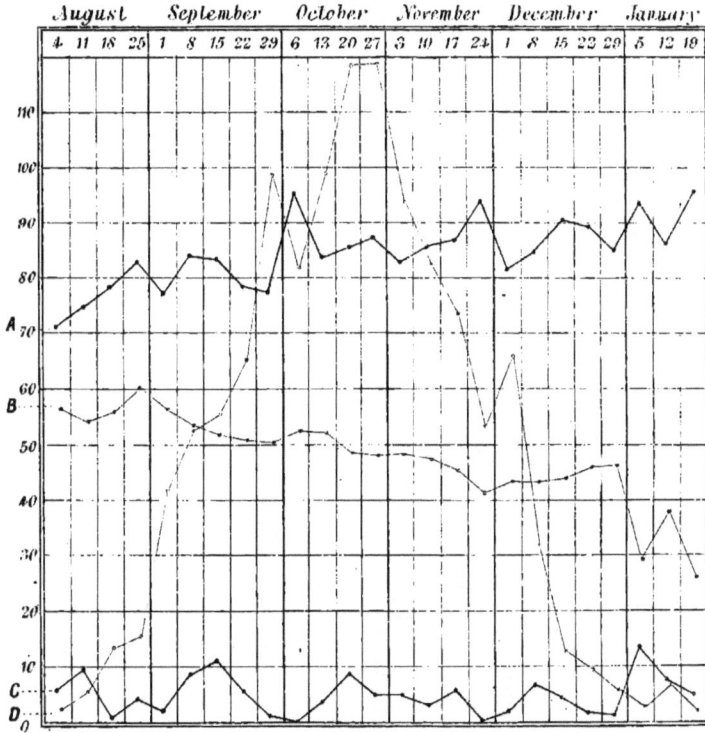

A. Percentage of Humidity;   C. Rainfall (10=1 inch of Rain)
B. Mean Temperature,   D. Cholera Deaths

on a fortnight of deficient rainfall. A gradual fall of temperature, a high humidity and a heavy rainfall then seemed to hold the epidemic in check. But the acme of mortality was reached in the middle of October, after a rise of 2° in mean temperature, and a very low rainfall. The increased mortality in the week ending October 13th is very remarkable when taken in connexion with the weather of the week before, which was continuously *calm, cloudy, foggy, damp*, *with a very high barometer, and a great deficiency of ozone* (the latter showing a mean value of only 10 per cent. at the Ordnance Survey Office, Phœnix Park).

The decrease in the mortality was consequent on a freshening and a change of wind from N.E. to S.W., a diminution of barometrical pressure, a moderate and continued rainfall, a rise in ozone to 70 per cent., and a gradually falling temperature. In the week ending December 1st we notice a final effort of the epidemic *after a week of low temperature.* But the influence of temperature seems, in this instance, to have been more than counterbalanced by deficient rainfall, a high barometer (pressure being a quarter of an inch higher than it had been for 3 or 4 weeks before), and the return of calm, damp weather, with a deficiency of ozone on three days. The coincidence of a high barometer with a great development of cholera has often been remarked, but striking exceptions are also on record. Keeping in view the fact that heavy rain and a strong breeze are most valuable detergents and disinfectants, I would suggest that the *calm weather* consequent on low barometrical gradients, so common in anticyclonic, or high-pressure systems, has more influence than the mere height of the barometer itself. In December the epidemic died out rapidly, and no death occurred later than the 29th of that month, on which day—it is most interesting to note—the intense frost of January, 1867, was ushered in by a fall of temperature amounting to 15° in a few hours.

That cholera tends to prevail in the warmer months of the year is sufficiently borne out by the history of the disease. In the accompanying Table are given the deaths from the disease by months in some of the great epidemics of late years, and the figures speak for themselves. In one case, that of Limerick, 1849, we meet with an early spring epidemic, and in January of the same year a large mortality from cholera occurred in England. But these exceptions only prove the rule. If from the totals we omit the Paris outburst of April, 1832, in which city the epidemic kindled into flame for the second time in July of that year, we have an increasing series of deaths from February to September, and a decreasing series from the last named month to December.

TABLE II.—Showing the DEATHS from CHOLERA, by MONTHS, in several EPIDEMICS, since 1832, in various CITIES and COUNTRIES of EUROPE.

MONTHS.	England, 1832.	England, 1849.	Paris, 1832.	Paris, 1849.	Dublin, 1849.	Limerick, 1849.	Dublin, 1866.
January, .	614	658	?	?	2	0	0
February,.	708	371	?	?	6	6	0
March, .	1,519	302	90	573	8	591	1
April, .	1,401	107	12,733	1,929	32	143	0
May, .	748	327	812	4,509	197	4	1
June, .	1,363	2,046	868	8,669	477	1	0
July, .	4,816	7,570	2,573	865	314	0	2
August, .	8,875	15,872	969	1,382	276	0	74
September,	5,479	20,379	357	1,142	298	1	270
October, .	4,080	4,654	62	115	49	0	508
November,	802	844	?	?	5	0	273
December,	140	163	?	?	0	0	66

MONTHS.	Sweden, 1834.	Sweden, 1850.	Sweden, 1866.	Christiania, 1833.	Christiania, 1850.	Christiania, 1853.	Christiania, 1866.	TOTALS.
January, .	?	?	0	0	0	0	0	1,274
February,.	?	?	0	0	0	0	0	1,091
March, .	?	?	0	0	0	0	0	3,084
April, .	?	?	0	0	0	0	0	16,345
May, .	?	?	0	0	0	0	0	6,358
June, .	?	?	4	0	0	0	0	13,628
July, .	30	0	483	0	0	6	0	16,659
August, .	5,904	209	943	0	0	164	8	34,676
September,	6,124	213	1,200	0	0	1,356	20	36,868
October, .	490	880	508	262	50	60	0	11,718
November,	58	342	30	538	37	11	0	2,940
December,	31	87	1	17	0	0	0	505

"Real epidemics of cholera,"

writes Professor Faye,*

" in the more rigorous season of winter have very seldom occurred, while sporadic cases have very frequently shown themselves even in winter. At Breslau a winter epidemic prevailed in 1848–49, continuing from October till March with the same fatality as had characterized summer epidemics at the same place ; and at Petersburg, as in several of the districts of Russia, cholera has prevailed in winter, although to a far less degree than in summer—so that the Russian physicians have often declared that the disease is prevalent in the winter quarter. At Bergen in Norway the epidemic of 1848–49 was also a winter epidemic. It is therefore not altogether without reason that cholera has been stated to observe no season, but if we take into consideration both the relative infrequency of its appearance in winter, and its impaired virulence under intense degrees of cold, this assertion as to the compatibility of the disease with a winter temperature experiences a very important limitation. Perhaps the explanation of the matter is not

* Loc. cit.

very remote. At Bergen, for example, the winter is often rainy and the air in proportion mild, so that the freezing of the earth's surface to any depth does not occur; and the winter of 1848–49 was really of this kind. It is well known also that cholera at Petersburg in winter time is almost exclusively confined to the unhealthy houses situated on the low and swampy banks of the Neva, belonging to an indigent labouring population; and indeed it is not strange that low-lying and overcrowded cellars, beneath which the soil has scarcely stiffened, with a favourable and confined oven-temperature, should foster the contagion, and occasion a constant though tardy propagation of the disease. Whether conditions of this kind held at Breslau I am unable to say, but in any case it is certain that violent epidemics during severe winter-frost very rarely, if indeed ever, occur."

Professor Faye goes on to say that, while the epidemic (of 1853) was at its worst at Christiania, the atmosphere was steadily warm and the air in addition clear and very still. This continued for about three weeks, during which the daily numbers of cases, which were then at the highest, scarcely varied. At this point of time—the middle of September—the air was set in motion by a strong and stormy north-west wind, and, remarkably enough, the number of cases fell *next day* to about one-half. Similarly, at Bergen, during the epidemic of 1848–49, a strong and cold north-easterly gale, supervening on a lengthened period of milder temperature, caused a considerable fall in the number of cholera cases.

B. In discussing the influence of season upon the progress of the recent epidemic of *small-pox*, we cannot, as with cholera, compare the meteorological conditions of the week before with the death-rate of any week. Small-pox has a definite period of incubation during which the disease lies dormant in the system, and it seldom kills before some days have elapsed from the earliest development of the symptoms. Making additional allowance for a few days delay in registration, I purpose to compare the deaths in a given week with the weather of three weeks before, and in doing this I am only following the precedent of all writers on the subject.

Small-pox is essentially a disease of winter and spring. In an accompanying Table I have entered the results of an analysis of most carefully compiled returns as to the prevalence of small-pox in Sweden during the eight years 1862–69 inclusive. The returns are extracted from exhaustive annual reports by Dr. Wistrand as to the morbility of Sweden, and are the direct fruit of an admirable system

E

of disease-registration which has been in operation for many years in Sweden, and also in the other Scandinavian countries. From these statistics it appears that the greatest prevalence of small-pox is observed in May, the cases in that month being 13·7 per cent. of the total cases occurring in the year; while the least prevalence is observed in September, when only 3·9 per cent. of all the cases in the year occur. From November the monthly number of cases is high, but from May a rapid decline in the prevalence of the disease takes place.

TABLE III.—Showing the prevalence of SMALL-POX in SWEDEN by MONTHS in the years 1862–69 inclusive.

YEAR.	Jan.	Feb.	Mar.	April.	May.	June.	July.	Aug.	Sept.	Oct.	Nov.	Dec.	TOTAL.
1862,.	28	55	43	40	56	53	66	67	46	42	25	52	573
1863,.	35	60	64	56	99	124	152	91	54	85	122	108	1,050
1864,.	163	148	157	222	267	314	187	130	160	168	244	317	2,477
1865,.	323	417	504	662	806	662	441	219	187	248	399	422	5,290
1866,.	647	484	498	636	619	530	315	184	89	101	254	159	4,516
1867,.	260	289	396	468	551	427	401	321	227	281	321	456	4,398
1868,.	595	544	619	813	770	567	471	228	211	186	373	608	5,985
1869,.	635	464	649	855	1,005	678	609	309	199	141	252	277	6,073
Totals	2,686	2,461	2,930	3,752	4,173	3,355	2,642	1,549	1,173	1,252	1,990	2,399	30,362

When due allowance has been made for difference of climate, these results agree very closely with the observations which have been recorded in this country on the relation of small-pox to season. A writer in the *Medical Times and Gazette* (March 11th, 1871) observes :—

"There is some reason for believing that the variations of the epidemic (of small-pox) from week to week are influenced to a certain extent by atmospheric conditions and more especially by variation in temperature."

He then quotes a series of remarkable coincidences between the fluctuations of mean temperature and those of the small-pox mortality in London during the winter of 1870–71. In the number of the same journal for May 13th, 1871, we read :—

"The epidemic has now lasted a good six months. It may be regarded as assuming a distinctly epidemic form in November, shortly after the mean temperature of the air had fallen decidedly below 50°. In the progress of the seasons we have now arrived at a time when this mean temperature is again reached. The mean temperature of the last three weeks, as recorded at Greenwich, has been 50°, 50·7°, and 49·7°. It is customary about the second week in May for some check in the consecutive weekly rises of temperature to take place, but after this in the ordinary or average

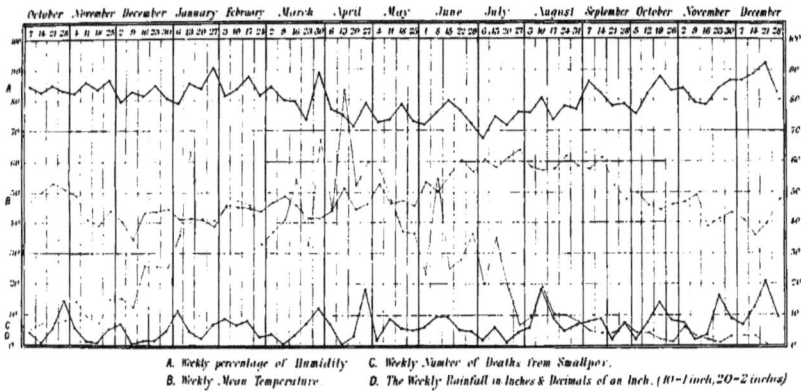

DIAGRAM IV.

Giving the Death Curve of the Small-pox epidemic of 1871-72, by weeks, in the Dublin Registration District, and the chief Meteorological Curves for the same period.

A. Weekly percentage of Humidity C. Weekly Number of Deaths from Smallpox.
B. Weekly Mean Temperature D. The Weekly Rainfall in Inches & Decimals of an Inch. (10=1 inch, 20=2 inches)

progress of events the steady rise towards the summer temperature may be expected to set in, and with it there is at least a hope that the epidemic will begin to fade."

A week later, the same writer, presumably, says :—

"The sudden fall of deaths in London from small-pox which occurred last week, namely, from 288 to 232, occurring about three weeks after the mean temperature of 50° was reached, appears to be confirmatory of the favourable hopes we expressed last week, that the epidemic had, for this season, arrived at its climax."

And so it had, for although the decline was occasionally interrupted, the virulence of the epidemic was broken in May, in accurate fulfilment of the anticipations which had been grounded on a consideration of the influence of temperature on its progress.

In Diagram IV. I have projected curves of the weekly deaths from small-pox during the recent epidemic in Dublin, and of the coincident meteorological conditions. It is much to be regretted that delays in registration have at times seriously marred the contour of the death-curve, which, I may mention, commences in the week ending October 7th, 1871, when an uninterrupted weekly mortality from small-pox began, and concludes in the week ending December 28th, 1872, when the first break occurred in the weekly death-return of the disease. Comparing the curves we find that a rise of mean temperature to 53° in the third week of October is followed by a decline in the mortality about three weeks latter. After the mean temperature had again fallen below 50°, the death-curve shows a tendency to rise, and a great development of the disease begins at the end of December, the mean temperature of the week ending the 9th of that month having fallen below 35°. The peak observable in the second week of January includes 36 deaths, which had only then been registered, although they had occurred in the South Dublin Union Workhouse since December 17th. Towards the end of February, a slight decrease in the deaths appears to be connected with the mild temperature of that month.

The peaks and depressions which follow are largely accounted for by irregular registration, but the greatest severity of the epidemic was experienced in the first half of April, a short time after a period of cold which was very intense for the time of year, snow and hail having fallen in large quantities (with keen north-easterly winds) on every day from the 21st to the 27th of March. The mean temperature of this period was scarcely 37°, or nearly 8° below the average. The number of deaths now began to decline, the mean temperature in two weeks (April 13th and May 4th) rising above 50°. The depression in the death-curve in the week ending June 1st, and the great rise in the following week depend on irregular registration. With the rise of mean temperature to between 55° and 60° in the middle of June, the weekly number of deaths falls permanently below thirty early in July.

It is interesting to note that abundant rainfalls seemed to be followed by remissions in the severity of the epidemic. A reference to the Diagram will show this very clearly; and the converse also, for the acme of the epidemic closely followed a period of comparatively dry weather and lower humidity.

But in connexion with the late epidemic, as regards Dublin, one of the most remarkable evidences of the dependence of the disease on climatic influences is found in the fact that in March, 1871, a well-marked tendency to an epidemic was noticeable. Local outbreaks of the disease took place in various parts of our city, and fatal cases occurred at Cork-street Fever Hospital. By the increasing temperature, however, the disease appeared to be held in check, notwithstanding the importation from England of many cases, until with the advancing autumn it blazed into an epidemic.

III. Influence of Season on (A.) Measles, (B.) Whooping-Cough, (C.) Scarlatina, and (D.) Fever.

I now pass to the third and last division of my subject, the influence of season upon our four principal endemic and epidemic diseases—endemic, alas! for their home is ever in the midst of us, and the graphic term "Fever-nest" is a significant and no less truthful recognition of the mournful fact. As briefly as possible I shall pass the four diseases in review in the following order:—(1) *measles,* (2) *whooping-cough,* (3) *scarlatina,* and (4) *fever.*

A. Following the death-curve from *measles* which I have projected in Diagram V., we notice (1) the periodical epidemic character assumed every second year or so by the disease, and (2) the remarkable tendency to prevail in the second and third quarters of the year which is shown by it. At present we have to do only with this last peculiarity. In epidemic years, on three occasions, the greatest mortality fell in the third quarter, and on one in the second quarter.

The non-epidemic years display an opposite tendency, the acme falling in the *first* quarter on four occasions. But in three of these instances, this acme was really only the dying-out of an epidemic (1866, 1868, and 1870). Practically then we may disregard these years, and we may look upon measles as essentially a disease of the spring and summer quarters.

Seeking for further light, I was led to analyse the weekly returns of deaths from measles during the 9 years, 1864–72 inclusive, in Dublin. As a result of the investigation I found that on an average the highest mortality fell in the twenty-eighth week of the year, and was 4·2 deaths—that from this period the average weekly number of deaths declined with slight oscillations to 0·6 in the fifty-first week, remaining very low until the twelfth week, when it again permanently reached two. Now the average mean temperature of the twenty-fifth week of the year for the 9 years under consideration was 58·6°. We may, therefore, conclude that a temperature higher than this is not favourable to the spread of an epidemic of measles. Similarly, the average mean temperature of the ninth week was 43·1°, while that of the forty-eighth week was 42·1°. As a low mortality from measles followed close upon the latter temperature, and lasted until the former temperature was reached in the early spring, the inference to be drawn is that a temperature below 42° is as unfavourable to the spread of the disease as a temperature above 59°. These results are in strict accordance with those arrived at by Dr. Ballard,* who says that the only condition concerned in the arrest of the spread of measles in summer is the rise of the temperature of the air above a mean of 60°, while towards winter a fall below 42° also distinctly tends to check the disease.

Proceeding from these results we see that the cold spring and summer of 1867 were especially favourable to the spread of the epidemic of that year, and that while, on the whole, the summers of *all* the epidemic years were comparatively cool, those of three non-epidemic years were hot and dry (1864, 1868, and 1870).

B. The consideration of *whooping-cough* need not delay us long. As was to be anticipated from the frequency of chest complications attending it, the disease invariably prevails most in winter, the greatest mortality generally falling in the first quarter of the year. Three epidemics of

---

* Eleventh Report of the Medical Officer of the Privy Council, 1868. No. 3, pp. 54–62.

whooping-cough occurred within the nine years 1864–72, and all of these reached their acme in January and February. It is curious to observe that the epidemics occurred in comparatively mild seasons, namely, those of 1866, 1868, and 1871. The epidemic of 1866 also was slow in dying out, as if the cold of the second quarter of that year had kept up the mortality. It would seem, indeed, that intense cold tended to check the disease, while moderate cold favoured its prevalence. And this view is borne out by an analysis of the weekly death-rate. The average weekly deaths numbered 5·9 in the second week—allowing then three weeks for (1) the period of incubation, (2) the length of the illness, and (3) the delay in registration, we find the average mean temperature of the fifty-first week in the nine years to have been 42°—a temperature which very remarkably corresponds with that of the three epidemic quarters 42·2°, and which is at least 3° above the average mean temperature of the coldest week in the year.

The lowest death-rate from whooping cough is met with in the twenty-eighth and twenty-ninth weeks, or about the middle and end of July, and accordingly, by a reference to Diagram V., we find the mortality to be, as a rule, lowest in the third quarter. The average mean temperature of the twenty-fifth week in the nine years was 58·6°, that of the twenty-fourth week having been 56·8°.

A rise then above this last temperature seems to favour the spread of the disease, although from Dr. Ballard's observations we learn that extremely high temperatures are inimical to the epidemic character of whooping-cough.

Besides the pronounced minimum of mortality from whooping-cough which I have described, another minimum falls in the twenty-first week, after which a recrudescence of the disease is observed for some weeks. With this temporary rise a low humidity, and a temperature of about 50° seem to be associated. Dr. Ballard especially mentions this June development of the disease as being most marked in Islington, and the Report of the Scottish Registrar-General for 1868 contains the following passages from the pen of Dr. Stark:—

"The first advent of really warm weather, during the past year, greatly increased the deaths from measles and whooping-cough; but the continuance of the warm weather rapidly diminished the mortality." "When the cold easterly winds began to blow in March, the deaths from whooping-cough in the eight towns, which numbered 87 during February, increased in March to 155, but under the influence of the spring weather fell to 145 during April, and to 131 deaths during May. During the high temperature of

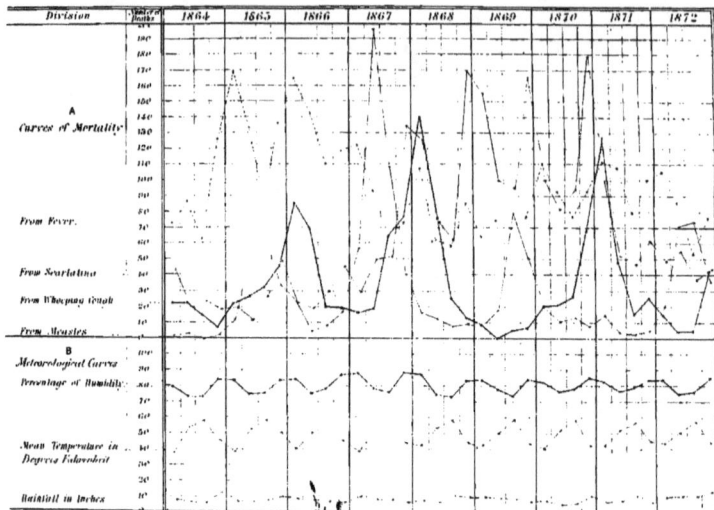

DIAGRAM V.

Compiled from the Quarterly Returns of Deaths in the Dublin Registration District for the past 9 years, and from the Quarterly Returns of Deaths within the Municipal Boundary of Dublin for the year 1864, illustrative of the effect of certain Meteorological Conditions on the Death Curves from four principal Endemic and Epidemic Diseases.

June, however, the deaths from whooping-cough rose to 165, the highest they had been during any month of the year; but instead of increasing during the much warmer months of July and August they rapidly fell, numbering 135 deaths in July, 121 in August, and 92 in September—the lowest number of deaths from whooping-cough during any month of the year." .

C.—"*Scarlatina*," observes the Registrar-General of England,* " discovers a uniform well-marked tendency to increase in the last six months, and attain its maximum in the December quarter, the earlier half of the following year witnessing a decrease." He illustrates this remark by a table, which is appended, showing the deaths in London from Scarlatina by quarters during four years.

TABLE IV.—Deaths in London from Scarlatina.

Years.	March Quarter.	June Quarter.	September Quarter.	December Quarter.	Total.
1861,. . .	420	326	467	1,145	2,358
1862,. . .	774	677	841	1,165	3,457
1863,. . .	880	1,055	1,519	1,621	5,075
1864,. . .	749	593	805	1,095	3,242
1865,. . .	566	–	–	–	–

From Dr. Wistrand's Reports on the Morbility in Sweden, I have compiled a table, which shows that scarlatina is as a rule most prevalent throughout that country in November, and least so in August, results which agree tolerably closely with observations in England, except as regards the September quarter.

TABLE V.—Showing the prevalence of Scarlatina in Sweden by Months in the years 1862–69 inclusive.

Years.	Jan.	Feb.	Mar.	April.	May.	June.	July.	Aug.	Sept.	Oct.	Nov.	Dec.	Total.
1862,.	371	246	194	177	88	93	99	157	134	95	103	116	1,878
1863,.	163	195	200	137	89	112	141	243	203	303	356	269	2,411
1864,.	303	369	307	312	203	247	192	220	217	202	376	563	3,511
1865,.	467	456	505	562	609	657	568	487	629	1,220	1,214	1,039	8,413
1866,.	1,122	930	763	802	948	818	781	706	659	863	970	962	10,324
1867,.	759	493	550	555	431	374	260	222	209	281	318	283	4,735
1868,.	273	215	233	235	218	205	263	188	260	386	564	644	3,704
1869,.	741	590	551	551	738	726	688	644	819	982	840	740	8,610
Totals	4,199	3,494	3,303	3,351	3,324	3,232	2,992	2,867	3,130	4,332	4,741	4,616	43,586

If we study the scarlatina death-curve in Diagram V., we shall be struck (1) by the remarkable epidemic period of 1867-70, and (2) by the great tendency displayed by the

* Twenty-eighth Annual Report of Births, Deaths, and Marriages, p. 38.

disease to prevail in the fourth quarter of the year. Even in non-epidemic years this tendency is generally noticed.

From an analysis of the weekly death-rate from scarlatina in Dublin during nine years, I have found that the disease was on the average most fatal in the forty-sixth week (8·2 deaths), and least fatal in the twenty-fourth week (1·9 deaths). The average mean temperature of the forty-third week was 47·9°, that of the twenty-first week was 52·1°. It would seem then that scarlatina shows a tendency to increase when the mean temperature rises much above 50°, while a fall of mean temperature below this point in autumn checks the further rise of the mortality.

Dr. Ballard draws inferences which confirm my results. He says :[*] —

"1. *That a mean atmospheric temperature of about* 60°, *or between* 56° *and* 60°, *is that most favourable to the outbreak of scarlatina.* 2. That for its free development it is necessary that the humidity of the atmosphere shall not much exceed 86, or be much less than 74. 3. That a higher temperature than 60° does not appear to be in itself unfavourable to the spread of scarlatina. 4. *That a fall of mean temperature below* 53° *tends to arrest an epidemic of the disease.*"

The remark made by this author as to the influence of humidity may explain the great dip in the mortality during the hot but *dry* summer of 1868.

But if a fall of temperature below 53° tends to arrest the disease, why is it that the mortality undoubtedly *continues high* during the colder winter months ? In Dublin it continues *very high* until the ninth week, and *high* until the nineteenth week. Here we are brought for the first time face to face with one of the most important factors in all sanitary problems, namely, overcrowding—or to give it a Greek dress—"ochlesis." We all know that scarlatina is not only one of the most contagious diseases in existence, but also that the *materies morbi* (whatever it may be), appears to be very easily diffused and remains in an active state for a lengthened period. Hence the difficulty of disinfecting the bed-chambers of scarlatina patients.

In these facts is contained the solution of our problem. As winter approaches, the instinct is to diminish the sources of ventilation, but among our poorer fellow citizens, badly clothed, and with inadequate supplies of fuel, unrestrained freedom is given to this instinct, but with most deplorable consequences. Every chink and crevice through which the outer air might gain access to the overcrowded tenements is eagerly sought out and effectually closed. And it is under these circumstances that scarlatina, favoured by the high and unwholesome temperature of the rooms, runs like wildfire among many families in the poorer parts of the city. But the mischief does not end here, for the contagious powers of the disease are called into full play, and so the richer and more affluent quarters of the city suffer in their turn from this dire pestilence.

We must not forget also that in winter the throat complications of scarlatina are likely to be more severe and more fatal than in summer and autumn.

D. *Fever* may well be described as both an epidemic and an endemic disease. In 1865 and 1866 it prevailed as an epidemic; it is however never absent from our city, and so its title to be considered an endemic disease also is established. Following the fever death-curve in Diagram V. we notice the remarkable tendency to prevail in the first quarter discovered by the disease. To this rule there are but two exceptions, one in 1864, the other in 1869. With regard to the former year, the increase in the second quarter may fairly be set down to (1) improved registration and (2) a commencing epidemic tendency. With regard to the latter year, we may remember the open mild character of the weather in the first two months, and the cold spring which followed. In fact, fever appears to depend especially on the weather. But in our consideration of fever it would be most desirable to isolate two forms of the disease—typhus and typhoid, or enteric, fever. With the former I have grouped the so-called simple continued fever, for I believe it to be more closely allied to typhus than to typhoid, and we must also remember that cases of simple fever are sometimes cases merely of typhus without any eruption—non-maculated typhus as it is termed.

Unfortunately, prior to the year 1869 no distinction between these two forms of fever was made in the Registrar-General's returns, but I have analysed the returns for the past four years in which such a distinction was made, and the results may be looked upon as at all events approximate to the truth.

In the first place, during the nine years, fever in general proved most fatal in the third and fourth weeks of the year (with 10·1 and 9·7 deaths on the average respectively), and least fatal in the 28th and 29th weeks (with 4·4 and 4·3 deaths on the average respectively), that is, the disease was most severe about the period of greatest cold, and least so early in July.

In Table VI., in which the year has been divided into thirteen periods of four weeks each, corresponding results are brought out.

TABLE VI.—Showing the MEAN NUMBER of DEATHS from FEVER, and the MEAN TEMPERATURE, in 13 periods of 4 weeks, during the years 1869–72.

No. of Period.	Mean Number of Deaths from Fever.	Mean Number of Deaths from Typhus.	Mean Number of Deaths from Typhoid.	Mean Temperature.	Per cent. of Typhus.	Per cent. of Typhoid.	Mean Fever Deaths, 1864–72.
I.,	34·4	19·2	15·2	40·5°	55·8	44·2	34·4
II.,	28·0	16·4	11·6	43·9	58·6	41·4	32·8
III.,	27·6	16·4	11·2	43·6	59·4	40·6	30·7
IV.,	27·5	17·0	10·5	47·0	61·8	38·2	27·9
V.,	32·5	17·5	15·0	49·4	53·8	46·2	31·6
VI.,	26·3	15·3	11·0	54·3	58·2	41·8	29·8
VII.,	22·3	11·5	10·8	59·3	51·5	48·5	20·9
VIII.,	20·5	12·5	8·0	61·0	61·0	39·0	20·7
IX.,	21·7	11·0	10·7	58·2	50·7	49·3	21·4
X.,	21·0	12·5	8·5	53·9	60·0	40·0	21·6
XI.,	27·5	14·2	13·3	48·1	51·6	48·4	25·7
XII.,	30·5	17·3	13·2	42·0	56·7	43·3	31·5
XIII.,	23·9	11·2	12·7	39·8	46·8	53·2	28·1
Average,	26·4	14·8	11·7	49·3	55·8	44·2	27·5

With respect to this and the following table it is necessary to explain that the fall in death-rate noticed in the thirteenth period, or last four weeks of the year, is apparently due to delay in registration at Christmas time.

TABLE VII.—Showing the MEAN NUMBER of DEATHS in DUBLIN from (1) MEASLES, (2) WHOOPING-COUGH, (3) SCARLATINA, and (4) FEVER ; and the MEAN TEMPERATURE, in 13 periods of 4 weeks, during the years 1864–72.

No. of Period.	Mean Deaths from Measles.	Mean Deaths from Whooping-Cough.	Mean Deaths from Scarlatina.	Mean Deaths from Fever.	Mean Temperature.
I., .	4·3	17·5	22·3	34·4	39·1°
II., .	5·1	15·4	21·4	32·8	41·7
III., .	6·8	11·0	16·0	30·7	41·4
IV., .	12·1	10·4	16·4	27·9	46·9
V., .	11·0	8·3	14·5	31·6	50·8
VI., .	12·5	7·0	12·5	29·8	54·9
VII., .	14·4	5·4	11·1	20·9	59·5
VIII., .	12·1	4·4	13·5	20·7	60·0
IX., .	9·7	7·8	13·7	21·4	58·6
X., .	8·7	6·7	18·5	21·6	54·2
XI., .	9·7	8·2	26·7	25·7	48·7
XII., .	5·0	11·4	26·4	31·5	43·5
XIII., .	4·2	11·2	22·8	28·1	43·4
Average, .	8·9	9·6	18·1	27·5	49·4

In Table VII. I have grouped the mean number of deaths from fever, as well as those of the other endemic diseases I have been considering, into thirteen periods of four weeks each. From this table it will be seen that fever becomes very fatal in autumn when the mean temperature falls below 54°, the mortality continues to rise with the falling temperature until January and February are past. Early in March the mortality declines, but rises again at the beginning of May, coincidently, it would seem, with a *lower humidity*. The decline is then very rapid, and the minimum is reached in the seventh and eighth periods, that is in July, and the first half of August. It is worthy of note that the sudden fall in the number of deaths in the seventh period follows the rise of mean temperature above 54° at an interval of some three or four weeks. Temperatures higher than 54° would, therefore, seem to have a controlling influence on the prevalence of fever, while temperatures below 54° seem to favour its development.

Table VI. shows the apparent influence of season on the two forms of fever—typhus and typhoid. The death-rate from typhus reaches a minimum in the ninth period, while the minimal death-rate from typhoid has already occurred in the eighth period, this form of fever exhibiting—as the summer rolls by—a decided tendency to increase at an earlier period than typhus. In this same table the calculated percentages of the two forms are also entered, and a striking increase in the per-centage amount of typhoid is noticed towards the close of the year. The highest per-centages of typhus are met with, on the contrary, in the seasons of winter, spring and early summer.

The reason for all this is not far to seek. Typhus is often intimately related to overcrowding, and bronchial or thoracic affections are amongst its most frequent complications. Hence we should expect to meet with it especially in the colder seasons. Typhoid fever, on the other hand, is connected with a specific contamination of air or water by sewage matter, and its secondary phenomena are developed generally in connexion with the digestive system. Hence, a great prevalence of this form of the disease was to be looked for in the warmer seasons, and more particularly at a time when the first autumn rains had washed into drinking wells and other sources of water supply the decomposing matters which had been innocuous so long as the skies were clear and the sun still high in the heavens. But this perhaps is only theoretical, and on the domain of theory I may not trench.

For this inadequate and imperfect sketch of a great subject
I ask your indulgence. Within the restricted limits of a
lecture, it would be impossible to do full justice to the
theme on which I have addressed you; indeed each separate
topic would claim a special lecture for itself. If, however,
some interest has been excited in a novel and all-important
inquiry my efforts will not have been made in vain. And
yet, all that we have considered to-day forms but one chap-
ter in the study of the relations between meteorology and
health. The influence of light on health, first shadowed
forth by our great countryman, Dr. Graves,* has been ably
handled by Dr. Forbes Winslow.† In the essay on "The
Influence of Light," by Dr. Graves, we meet with this touch-
ing and eloquent passage:—

"I need not observe that the flowers and leaves of all plants
court the light; indeed this tendency is manifested sometimes in
a very curious manner. This is exemplified in the various flowers
which adorn the dark and comfortless abodes of the tradesmen in
the Liberties of Dublin. These poor creatures (for however poor
the being is, or however confined by the nature of his employment,
he never forgets the green freshness and living loveliness of nature)
delight in flowers and birds, and in their windows will frequently
be seen a geranium, almost as sickly as its owner, turning its lank
and stunted leaves with unvarying constancy towards the light."

„Die Pflanze selbst kehrt freudig sich zum Lichte."
SCHILLER—"_William Tell._"

Among other cognate inquiries I would particularize only
the works of Dr. Angus Smith on "Air and Rain," of Dr. Cor-
nelius Fox on "Ozone and Antozone," and of Professor Buhl,
of Munich, on the "Relation between Typhoid Fever and
the Height of the Underground or Subsoil Water."

When, indeed, we reflect on the vast extent of the subject
of Meteorology and Health, we are ready to exclaim with
the great Physician of Antiquity—

"Life is short, and Art long; Occasion fleeting;
Experiment dangerous, and Judgment difficult." ‡

But in conclusion, one word as to the practical bearing
of our investigations. It would seem that some of the
diseases we have been considering tend to prevail in the
warm seasons of the year, others in the cooler seasons.
There is, further, only too good reason for believing that
the mortality of many diseases included even in the former

---

* "Studies in Physiology and Medicine," p. 26.
† "Light: its Influence on Health." 1872.
‡ "Ὁ βίος βράχυς, ἡ δὲ τέχνη μακρή· ὁ δὲ καιρὸς, ὀξύς· ἡ δὲ πεῖρα,
σφαλερή· ἡ δὲ κρίσις χαλεπή."—HIPPOCRATES—" _Aphorisms._"

class is increased by overcrowding and its attendant evils in winter, while the influence for evil of this flagrant breach of sanitary law on winter maladies is almost beyond belief. Overcrowding, alas! is but another name for poverty. And poverty means—want of fuel, deficient food, deficient clothing. What an argument have we here for the promoters of coal funds, soup kitchens, and clothing clubs? If it be true that cold gives rise to bronchitis, inflammation of the lungs, pleurisy, and a host of other "ills that human flesh is heir to," is it not incumbent upon us, sanitary reformers, and pioneers of *Preventive* Medicine, to obviate so far as lies in our power the evil effects of cold? In the case of scarlatina, again, and other infectious diseases, let refuges be provided to which we may remove from the sick-room the still healthy members of a family stricken by the disease. As regards the maladies of summer, too, we may do much. The providing of wholesome food, the interdicting of unripe fruit and putrid vegetables, the free use of suitable disinfectants in sewers and latrines, and above all, a pure water supply, such as we already possess, will have the happiest results.

But of all the diseases, typhus fever is, perhaps, the most preventable, depending as it does so largely on overcrowding and bad ventilation. How are we to deal with it from a preventive point of view? Let us hear the Registrar-General of England* on this point—

"Fire is a necessity of life in this climate, and a warm hearth mitigates the severity of winter. Fire is as much required by the poor as by the rich, and a tax on coals like a tax on salt presses with undue severity on people of small means."

And so it is. Our poorer fellow-citizens have to do battle with snow and ice, hail and tempest. Their weapons of defence in this otherwise unequal warfare must be raiment, food, and warmth. Lo! there on the journey of life lies the wounded, the helpless wayfarer, cold, and naked, and hungry—be you the good Samaritans.

* Twenty-seventh Annual Report, for 1864.

LECTURE IV.

ON THE GEOGRAPHICAL DISTRIBUTION OF DISEASE.

By JAMES LITTLE, M.D.,

Professor of the Practice of Medicine in the Royal College of Surgeons.

In dealing with the subject of the Geographical Distribution of Disease it is desirable to keep in view the two great classes of disease. In the admirable lectures of Dr. Stokes and Dr. Moore you have already heard a good deal about one of these classes, namely, *Zymotic* Diseases. The derivation of this term almost suffices to explain the sense in which we apply it. It comes from a Greek word which signifies "leaven," and under the term are grouped a number of diseases in which the disease-process is supposed to bear an analogy to that which is set a-going when leaven is added to dough. One of the most prominent members of the zymotic group is small-pox. When a certain quantity of the *contagium* of small-pox is introduced into the system nothing is observed for some time, but after the lapse of a certain number of days, varying from nine to twelve, according to the mode in which the poison is introduced into the system, changes begin in the individual—he becomes ill, various symptoms are developed, and there is a multiplication of the poison in his body. One of the essential characters of these zymotic diseases is that the poisons which produce them are all received from *without*—the individual gets the poison from some source external to his body.

The other class of disease is the *Diathetic*. This term is derived from a Greek word which signifies "disposition." There exists in the bodies of some people a disposition or predisposition to a certain disease, or more correctly to a certain kind of vital action which results in disease, and an essential feature of this class is that the cause of the disease is generated *within* the system of the sufferer, and is not received from without, as is the case in diseases of the zymotic class. Gout and scrofula are examples of diathetic diseases. Men do not exhibit the phenomena of gout or scrofula because some poison has gained access to their bodies from without, but because there is something wrong in the chemico-vital changes going on in their own bodies, because in them there is a disposition or predisposition to form an unhealthy kind of blood and tissue. Diathesis is sometimes hereditary and sometimes acquired, but whether hereditary or acquired it may be intensified or lessened by the habits of the individual and the circumstances in which he is placed.

Some diseases belonging to both these groups prevail in every part of the world, or nearly in every part of the world, but the majority are confined to certain areas.

I. Some are confined to a very limited area, beyond which they are never, or very seldom found ; for instance, along a certain part of the coast of Hindostan and the opposite coast of Ceylon, between 15° and 20° north latitude, and extending not more than sixty miles inland, there prevails a constitutional disease known as Beriberi. In persons affected by it the blood becomes watery, and ultimately dropsy occurs ; it prevails principally in gaols and overcrowded barracks. In this district many conditions unfavourable to health exist ; we do not know the precise ones which determine the occurrence of Beriberi, but we do know that by attention to ventilation, the use of pure water, the avoidance of damp and cold, and the administration of small quantities of iron, persons living within the affected area may be protected from the malady.

II. Other diseases, again, though usually confined to a limited area, extend beyond it under exceptional meteorological conditions. For instance, in the equatorial region of America, extending from 48° north latitude to 35° south latitude, Yellow Fever prevails. In some part of this region it is always present, but it requires a temperature of 72° Fahr. for its propagation, and hence, although there is constant communication between the countries in which the disease prevails and England, and ships have several times arrived in British ports with the disease, it has never spread in these islands ; it has spread, however, when introduced into ports in southern Europe, where the temperature and other circumstances were favourable to the reproduction and diffusion of the poison on which it depends.

III. In Cholera, again, we have an instance of a disease which has for some time prevailed in nearly every part of the world, and under very varied climatic conditions, but which nevertheless has a limited area of *persistent existence* from which its epidemic journeys commence, and where, after it has died out in other lands, it continues to prevail.

IV. In Influenza we have an example of a disease which has no persistent area, but which, arising sometimes in one part of the world and sometimes in another, spreads thence with great rapidity.

V. Finally we have diseases such as Goitre and Leprosy, which are met with in nearly every country in the Old and New World, but only in limited districts of these countries, where certain unhealthy influences generate and perpetuate them. Goitre is found in India, in the New World, in parts of England, and in Switzerland, but always where the water used for drinking contains lime in considerable quantity. Leprosy is a terrible disease, which once existed here. It prevailed in various parts of Great Britain and Ireland, and up to the end of the last century in the Shetland Islands; at present it prevails in India, in North America, in Equatorial Africa, and elsewhere, but whether in the warm or cold regions, in the Old World or the New, it always presents the same characteristics. It is an hereditary constitutional disease, prevailing among people who are badly fed, and live in filth and misery.

If, setting aside the two last classes, we proceed to inquire into the circumstances which cause the limitation of certain diseases to certain areas, we will find that temperature is the most powerful—so powerful is it found to be that some years ago Mr. Keith Johnston, the eminent geographer, constructed a map in which he divided the world into three great disease realms, according to isothermal lines; that is, according to lines passing through places of mean annual temperature. If we look at Mr. Johnston's map we will find that the isothermal lines do not run parallel with the equator; nearness or distance from the equator is indeed the circumstance which has most effect in determining the temperature, but other circumstances often interfere to disturb the influence of this condition, and hence the lines of mean annual temperature at some points approach and at some recede from the equator.

The TORRID Disease Realm has its centre at the equator, where the annual mean temperature is 82° Fahr., and extends north and south to the isothermal line of 77° Fahr. in it therefore we find the Southern States of North America, Mexico, and the northern part of South America, the great region of Central Africa, Arabia, India, and China. This realm includes the most unhealthy portions of our globe. Most of its diseases are of the zymotic class, and in their symptoms resemble the affections which prevail among ourselves during the heat of summer and autumn. Over the entire of this disease-realm intermittent fevers prevail. In the portion of it which lies in the New World yellow fever is the most wide-spread and fatal malady, while in that which lies in the Old World cholera is the most important.

Intermittent and remittent fevers are also frequently spoken of as malarious fevers, because they are supposed to depend on an emanation from the soil, to which the term malaria has been applied. The fevers which prevail in this country are called *continued* fevers, because in them from the commencement of the illness until its termination there is persistent fever, that is, persistent elevation of the temperature of the body above the limits of health ; in intermittent fevers, on the contrary, after the patient has suffered for some hours, the fever subsides and does not return until the following day or until the second day. Some intermittent fevers are comparatively mild while some are rapidly fatal, but whether the fever has been a mild or a severe one the person who has suffered from it—instead of being protected from a second attack, as is the case after recovery from our continued fevers—has a permanent impression made on his constitution, so that for years after he now and then shows symptoms of the febrile paroxysm. Though it is within the tropics that these fevers are most prevalent, they are seen in a severe form in Southern Europe, and in a less severe form in Britain, and wherever they prevail a certain condition of soil exists, namely, an alluvial soil, which, never thoroughly dry and never completely flooded, is daily exposed to a hot sun ; heavy rains on the one hand and long continued drought on the other prevent the development of malaria, and it is most intense during the autumn when the damp ground is covered by decaying vegetable matter. In the deltas of great rivers, where an immense body of water finds its way to the ocean through innumerable small streams, as is the case with the Ganges and the Nile, malaria is intense ; it also prevails where extensive systems of irrigation are carried out, and where there are undrained swamps. Some of those who are now present have no doubt visited the beautiful Basilica of St. Paul, a few miles from Rome, and have heard how the monks who have charge of it are obliged each summer to leave it on account of the fatal fever which is generated in the undrained Campagna, yet this region was at one time quite healthy, when the aqueducts and watercourses, which are now in ruins, carried off the water and kept the soil dry.

I have already mentioned a few of the characteristics of yellow fever, the most conspicuous disease found in the

F

torrid realm in the New World, but as cholera, the prominent disease in the torrid realm in the Old World, possesses a far greater interest for us, I shall devote to it a little more attention.

Cholera,* as all here are aware, has from time to time prevailed over nearly all the world, but in most countries after raging for some time it dies out, and is no more heard of for years ; but there is one part of the earth's surface where this is not the case. Somewhat similarly typhus fever has under certain conditions, which I am sure you will hear fully described by my friend Dr. Grimshaw, prevailed extensively in many parts of Great Britain and of the Continent, but in most of these situations it dies out as soon as the crowded camps or other aggregations of human beings among whom it has prevailed have been broken up, but in this country, and more especially perhaps in this city, typhus fever in isolated cases is constantly present, and under certain favouring circumstances becomes epidemic. Cholera has in Lower Bengal, and especially in the great towns of Dacca and Calcutta, a home ; when its ravages have ceased elsewhere, it still prevails in this region, sometimes appearing in isolated cases, and sometimes putting on the character of an epidemic. Here it seems to find the conditions necessary for its permanent existence or constant reproduction, or perhaps I should rather say in Lower Bengal there are circumstances which prevent the disease dying out as it does elsewhere. In the great Delta of the Ganges we find everything which is necessary for the development of malaria in its greatest intensity, an alluvial plain so flat, that for 200 miles inland it barely rises above the sea level, exuberant vegetation, vast expanses of jungle, a great network of rivers and canals, and a tropical sun ; in the habits of the people too we find everything favourable to the spread of an epidemic. "A bustee or native village," says Dr. Tonneore, "generally consists of a mass of huts, constructed without any plan or arrangement, without roads, without drains, ill-ventilated, and never cleaned. Most of the villages and towns are the abodes of misery, vice, and filth, and the nurseries of sickness and disease. In these bustees abound green and slimy stagnant ponds, full of putrid vegetable and animal matter in a state of decomposition, whose bubbling surfaces exhale, under a tropical sun, noxious gases, poisoning the atmo-

* The most complete account of cholera, and especially of the routes by which its epidemics have travelled, will be found in "A Treatise on Asiatic Cholera," by C. Macnamara, Surgeon to the Calcutta Ophthalmic Hospital. London: Churchill. From this work the accompanying Map is, with certain alterations, borrowed.

sphere, and spreading around disease and death. These ponds supply the natives with water for domestic purposes, and are also the receptacles for their filth. The arteries which feed these tanks are the drains that ramify over the village, and carry out the sewage of the huts into them. Their position is marked by a development of rank vegetation.

" The entrances to these bustees are many, but not easily discoverable, while the paths are so narrow and tortuous that it is difficult for a stranger to find his way through them. The huts are huddled together in masses, and pushed to the very edge of the ponds, their projecting eaves often meeting together, while the intervening spaces, impervious to the rays of the sun, are converted into necessaries, and used by both sexes in common. In these huts often live entire families, the members of a hut all occupying the single apartment of which it is not unfrequently composed, and in which they cook, eat, and sleep together, the wet and spongy floor, with a mat spread on it, serving as a bed for the whole.

" The distinction of caste extends to these bustees; but it assumes in these places a new form, by the fact that some portion of them, called parrahs, are inhabited by people of one occupation or trade, whose habits of living give a distinctive feature to each parrah, and modify its general appearance. Amongst the Hindoos the worst and filthiest bustees are those occupied by Gowallahs, Coloos, Chumars or Moochees. Amongst Mahomedans the worst and filthiest bustees are those occupied by Garrywans and Khollasees. In bustees occupied by Gowallahs, in addition to the usual filthy tank, the water of which is used by them to dilute the milk sold for public consumption, there are pools of liquid filth covering a large surface, the area of one of them I have ascertained by actual measurement to be over 150,000 square feet.

" None of these villages possesses a single road, or thoroughfare properly so called, through which a conservancy cart or even a wheelbarrow can pass in order to remove the filth. This filth is laid at the door of every hut, or thrown into a neighbouring cesspool. Not a single hut in the village is properly built. The dwellings are badly constructed, crowded together without regard to ventilation or the means of being kept clean. The principal defects are due not only to ignorance and utter disregard of all sanitary considerations by the ryots, but also to the apathy and negligence of the impropriators, who care very little about the welfare of their tenants provided that their rents are paid regularly."

This is a faithful and not overdrawn description of village life in Bengal, and it presents to us conditions eminently favourable to the prevalence of such a malady as cholera. Observation at home has shown that some diseases, markedly typhoid fever and cholera, are very frequently spread by the emanations from the sick finding their way into the drinking water, and thus gaining access to the bodies of the healthy. Now, the drains which Dr. Tonneore describes, though usually stagnant and invariably receptacles for the refuse of the huts between which they flow, during heavy rain are flushed, and discharge their contents into the neighbouring tanks from which the drinking water of the village is drawn. Need we be surprised then that in Lower Bengal cholera never dies out ? and it is from Lower Bengal that it invariably sets out on its epidemic journeys. The routes it has followed on these journeys have been various, but the more closely we examine them the more convinced will we be of the truth of the doctrine taught long ago by the late Dr. Graves, that cholera travels along the highways of human intercourse, carried from city to city and from country to country, as the case may be, in the caravan, in the crowded ship, in the coach in the olden time, or in the railway carriage in our own day. Before us we have a map on which is shown the route by which it reached Ireland on its last visitation.

In the beginning of the year 1864 every part of the world was free from cholera except Bengal. The disease had died out except in its endemic area, but during that year it spread through Central India and into the Bombay Presidency. There are several places in India to which, at certain seasons, vast numbers of the natives journey from great distances; at some of them fairs are held, to others they go on pilgrimages. Into these assemblages cholera was introduced by those who had come from Lower Bengal, and when the gatherings had broken up the individuals who had contracted the disease carried it with them to their native towns, and so it was found that in 1865 cholera prevailed as a severe epidemic in Bombay. Between Bombay and the Southern Coast of Arabia there is constant commercial intercourse, and in March, 1865, cholera was raging at Mokalla and Mocha, to which it had been carried by native traders from Bombay. Now, this year, 1865, was one of special importance in the Mahometan calendar, and during the early summer devout Mahometans set out from every country in which that faith prevails on a pilgrimage to Mecca and Medina; they came from the East, from India

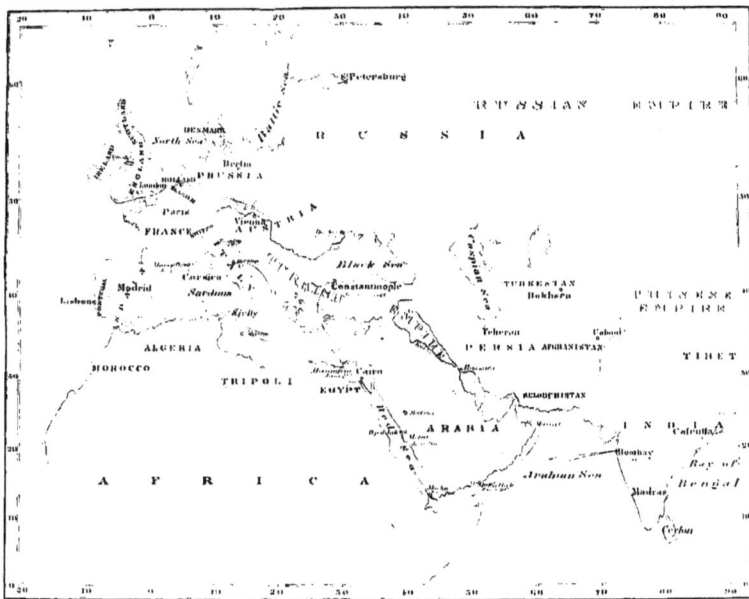

and China, and the Indian Archipelago ; they came from the
West, from Turkey, and Egypt, and assembled to the number
of 90,000 around the holy cities. Among the vessels which
conveyed them were two, the *Persia* and the *North Wind.*
These ships came with pilgrims from Singapore, and after
calling at Mokalla disembarked their passengers at Djeddah,
the port of Mecca; but between Mokalla and Djeddah
cholera appeared among the passengers, and with them it
found entrance into the assemblage of pilgrims around
Mecca. These unfortunate people were living under conditions in the highest degree favourable to the spread of
any pestilence ; they were crowded together, very scantily
supplied with water, and surrounded by every kind of filth,
and by the remains of the animals which had been sacri-
ficially slaughtered. It is believed that 30,000 persons
died, and those who were left hastened away from the plague-
stricken encampment. As I have already mentioned, many
of the pilgrims came from Turkey and Egypt, and on May
19th the first ship with returning pilgrims reached Suez ;
many of the passengers had died of cholera during the
voyage, the others were immediately conveyed by rail across
the isthmus to Alexandria, where they encamped outside the
town. The condition of Alexandria and its environs at this
season was eminently favourable to the multiplication and
diffusion of the *contagium* of cholera. In a less degree Lower
Egypt presents the same features as Lower Bengal. It is the
delta of a great river ; it is exposed to an almost tropical
sun; its vegetation is luxuriant and its inhabitants are un-
cleanly. Here cholera broke out and prevailed with great
severity. The port of Alexandria too is the place of all
others in which the greatest facilities exist for the rapid
diffusion of any pestilence. In it you may see every day the
flag of almost every European nation. It is the great high-
way between Europe and the East, and from it are daily
starting English, French, Russian, and Austrian steamers.
The first case of cholera occurred at Alexandria on June 2 ;
on June 11 it was at Marseilles, on June 28 at Constantinople, and on July 7 at Ancona. Considerable precautions
were adopted to prevent its entrance into English ports, and
it did not appear at Southampton until September 17 ; it
did not spread there nor did it come to Ireland by that
route. We have now reached the autumn of 1865. During
the autumn and the ensuing winter it prevailed in various
parts of northern Europe; it had found its way there by two
roads ; from Constantinople it had spread along the shores of
the Black Sea and up the Danube, and by a totally different

path it had gained access to the very heart of Europe. Not a few of the pilgrims who fled from Mecca were Persians, and they journeyed homewards along the coast of Arabia and up the Persian Gulf; they carried the cholera with them up the Gulf, along the valley of the Euphrates, to the shores of the Caspian Sea, and hence it was soon carried into Russia. Reaching northern Europe by both these routes in the spring of 1866 it prevailed extensively in Holland, and on May 2 two persons who had just arrived from Rotterdam died of cholera in Liverpool; and, beginning in the streets inhabited by the Dutch and other foreign emigrants and sailors, cholera spread through various parts of Liverpool. On July 26 a girl died of the disease in Dublin, and on investigation of the case by Dr. Mapother it was discovered that she came from a Liverpool lodging-house frequented by these sailors.

The ARCTIC Disease Realm extends from the isothermal line of 41° Fahr. to the poles. In the countries lying in this region there exist many conditions unfavourable to health; extreme lowness of temperature and dampness are among the most powerful; in this region vegetation is less rapid and luxuriant, and the great body of the people have a less abundant supply of food; nor must we omit the long nights and short days and deficient sunlight from which those countries which approach the poles suffer.

The lowness of temperature indeed is not an unmixed evil, for it is eminently unfavourable to the spread of zymotic diseases, which, in consequence, seldom prevail in the Arctic realm. It is from constitutional diseases that the inhabitants of countries within this realm chiefly suffer; the conditions under which they live are unfavourable to long life or vigorous health, and they suffer in consequence from scrofulous diseases, and other diathetic diseases characterized by a low standard of vital action.

Between the torrid and the arctic disease realms lies the TEMPERATE, in which our own land is situated. The diseases met with in this realm will be largely discussed, I have no doubt, in the subsequent lectures of this course; they are due less to the climate in which the inhabitants of this region live than to their habits—to the artificial life they lead and the various unfavourable conditions imposed upon them by the crowding and the struggling for existence in our great towns, and the still more unfavourable influence exercised by their vices. The deadly miasm of the tropical jungle does not kill more surely, the bad food and sunless winter of the polar regions do not lower the standard of health and shorten life more certainly than does intemperance, which, in these lands, at the present day, prevailing as it does, among rich and poor, seems likely to inflict a blow on England's greatness more deadly by far than any we could receive from pestilence or war. Zymotic and diathetic diseases are to be met with in about equal proportions in this the temperate disease realm; among the former we have typhus fever, the crowd-fever of the poor, and typhoid fever—a malady which seems very intimately connected with the extensive system of sewerage which the arrangements of our modern houses require,. and is hence as commonly, nay more commonly, met with in the mansions of the rich and the comfortable homes of the middle classes than among the very poor; it is the fever which is endemic in many of the large continental towns, such as Naples and Geneva, and which in these cities not unfrequently causes the death of English tourists, who, coming freshly under the influence of the miasm, are specially liable to be affected by it. Conspicuous among the diathetic diseases we have consumption—a malady which is the expression of a low state of vitality, however induced, but which recent observation has shown to be in a great degree preventable; investigations carried on in England, under the direction of the medical officer of the Privy Council, have brought to light the fact that the mortality from consumption stands in a very constant relation to the dampness of the soil, and that where effective drainage has diminished this the mortality from comsumption has, in proportion, been lessened. In Salisbury the consumption death-rate has diminished one-half since the town was thoroughly drained.

The subsequent lectures of this course will, if I mistake not, impress strongly upon you this fact, that it is not only the duty but the interest of the rich to care for and to seek the health of their poorer neighbours; you will learn how the fever which is bred in the lanes and alleys of our city creeps into the broad streets and open squares, and the facts I have mentioned teach the same lesson. I have shown you how the pestilence, which finds the conditions necessary for its permanent existence in the undrained swamps and filthy villages of England's great dependency, travels thence until it reaches the busy hives of English industry and gathers its victims from the mansions and homesteads of Britain itself.

LECTURE V.

ON ZYMOTIC AND PREVENTABLE DISEASES.

By THOMAS W. GRIMSHAW, Esq., M.D.,

Fellow and Censor of the King and Queen's College of Physicians in Ireland; Physician
to Stevens' Hospital, and the Fever Hospital, Cork-street, Dublin.

THE subject which I have selected for this lecture is a large
one—so large that it is scarcely capable of being treated
with sufficient depth to make it interesting without being
made dull or abstruse from the close condensation of the
large number of facts, theories, and suggestions which I shall
have to pass in review in the short space of time at my
disposal. Fortunately, those who have gone before have
cleared the way for me, and I have no doubt that those who
will follow in this course of lectures will fill up the many
vacancies which I shall have to leave in this discourse. The
title of this lecture is "Zymotic and Preventable Diseases";
and I have chosen it with a view of indicating, as closely
as a title possibly can, the nature of the subject I have to
refer to. I do not propose to discuss all the diseases which
might be included among the preventable, or more correctly,
the controllable class, but only such as come under the head
of "Zymotics," and therefore more immediately affected
by public measures of prevention, and by the conditions
which affect large communities. A certain class of diseases,
namely, those owing to unhealthy trades are certainly controllable, but will be discussed in the lecture to be delivered
by Dr. Mapother. In the first place, I will give you an idea
of what I consider to be a preventable disease, and with that
view I shall define it as any disease which arises or spreads
in consequence of the wilful or careless violation of the laws
of nature, which we know it is necessary to observe to insure
the preservation of health or prevent the spread of disease.
I must also point out the nature of some of these laws of
health, and show the result of their violation. Some persons
have objected to the term "preventable" being applied to
any disease, believing conscientiously that diseases are a
direct visitation of God upon his people for his own wise
purposes and their benefit. I most heartily concur in this
opinion, for if ever direct judgments fall upon mortals for

sins committed, there can be no better examples than those derived from the terrible disasters which have so often followed the wilful or careless violation of the sanitary laws of nature, or, as I prefer to call them, the sanitary laws of God. I say, therefore, with all humility, that the term preventable, as applied to diseases, is in no way impious. Unfortunately, those diseases which are produced by personal and private bad habits are beyond the immediate control of public measures, and are outside the scope of this lecture.

Of these diseases Dr. William Farr remarks :—

"It is here that the various forms of plague are found which experience has shown are influenced to a large extent by sanitary conditions, thus small-pox is diminished by vaccination, enteric fever by sweetness of air, typhus by ventilation, erysipelas (St. Anthony's fire) by cleanliness, metria (puerperal fever) by isolation of mothers, diarrhœa and cholera by the exclusion of sewage from waters in domestic use."

Now, what do I mean by Zymotic diseases ? The term Zymotic has been adopted by those who believe that in these diseases a peculiar pathological process goes on allied to if not identical in nature with fermentation, as observed outside living bodies, this fermentation results in the production of that peculiar train of symptoms characteristic of each one of these diseases. I cannot here enter into the proof or disproof of this theory of Zymosis ; but I may state that I believe it to be the true foundation of the pathology of these diseases, and, up to the present, the only one which offers to my mind a reasonable explanation of the phenomena which attend their development, propagation, and results.

I use the word Zymotic only as a term used by the Registrar-General in his returns to signify those diseases commonly known as contagious febrile affections, together with a few other forms, which are not usually accompanied with febrile symptoms, and a few others which may not be, or are but slightly contagious. The chief diseases of this class are Fevers, Diarrhœa, Scarlatina, Small-pox, Whooping-cough, Cholera, Measles, Erysipelas, Metria (or Puerperal Fever), Croup, Diphtheria. There are some others less important, but these are sufficient to show the nature of the class, and their names are, I fear, but too familiar to most of you. We must consider them from several points of view :—

1st. The damage they inflict upon us.
2nd. The conditions under which they spread.
3rd. The conditions under which they arise.
4th. The means suggested for their control.

First,—The amount of danger done to us is immense. Thus from a return ordered by the House of Commons, on the motion of Mr. W. H. Smith, M.P. for Westminster, we find that of the 3,249,077 deaths which occurred in the United Kingdom during the five years 1865 to 1869 inclusive, 712,277, or 21·9 per cent. were caused by Zymotic diseases, in other words, about 1 in every 5 of the deaths is caused by a disease of this kind, and we lose at the rate of about 150,000 people annually in the United Kingdom from Zymotic diseases. The deaths were distributed between the three countries in the following proportions :—

England—111,418, being ¼ of the total mortality, or 1 in 190 of the population.

Scotland—16,193, being ¼ of the total mortality, or 1 in 194 of the population.

Ireland—18,416, being ⅙ of the total mortality, or 1 in 308 of the population.

From this we see that Ireland, with all her sanitary defects, is better off than the sister countries in this respect, and I believe this is owing to the energetic efforts of the well organized though badly paid system of Poor Law Medical Service, of which we have had so long to boast of in this country. When to this organization is committed the duty of preventing as well as curing disease, we may expect a still further reduction in the mortality from these diseases.

The deaths in Ireland from the principal Zymotics in the five years thus mentioned were—

Fever, . . .	21,895,	or at the rate of 4,379	per annum.
Scarlatina, .	16,474	„ „	3,295	„
Diarrhœa, . .	10,081	„ „	2,016	„
Whooping-cough,	9,475	„ „	1,895	„
Small-pox, .	1,553	„ „	314	„

From this table it appears that by far the most serious of these Zymotics are such as are always amongst us— 1st, Fevers, 2nd, Scarlatina. I shall presently show more fully that the constantly present Zymotics are far more serious than those which only annually visit us, such as Cholera, Small-pox. While these latter quickly sweep away large numbers, thereby striking us with terror, the former gradually and silently eat away thousands without much notice being taken of the effect produced by them. Let us see how these epidemics affect our city of Dublin. The total number of deaths in Dublin since the Registration Act came into force in 1864, up to the end of last year, 1872, was

73,661, or a weekly average of 157; of these, Zymotic diseases caused 17,156, being 22·6 per cent., and equal to a weekly average of 36·5. These were distributed as follows among the various Zymotics in order of their numbers :—

	Total.	Rate per annum.
Fevers,	3,506	389
Diarrhœa,	2,576	286
Scarlatina,	2,407	267
Small-pox,	1,699	188
Whooping-cough,	1,464	162
Cholera,	1,293	143
Measles,	1,124	124

The balance is made up of several others less destructive, but probably not less controllable (see Diagram IV.)

This is farther shown by a comparison of the admissions into the Dublin Fever Hospitals, as seen in Table I. and accompanying diagram. These hospitals take in all kinds of Zymotic diseases, and although a number of other acute diseases not of the Zymotic class gain admission, yet those may be fairly set against the considerable number of Zymotics admitted into the general hospitals, so that Table I. fairly measures the prevalence of Zymotic diseases in Dublin for the 15 years ending March 31st, 1871; but of course only represents a portion of the cases, perhaps a large portion of the worst cases, but scarcely any of the mild ones. The great bulk of the cases here tabulated in Table I. consist of Fever. On inspecting the table, it will be seen that, commencing with the year 1857, when the Board of Superintendence furnished the first Report, the admissions fell until the year 1859, when they rose again for one year (1860), then fell again for one year (1861), to rise again continuously until 1866, when the number of admissions reached 3,562; in this year cholera also prevailed, and that disease is included in the Hardwicke Hospital Returns to the number of 187. Cholera was not admitted into Cork-street Hospital.

Zymotics, especially fever, prevailed to a greater extent in the year 1866 than it has done at any time during the period under consideration; the numbers in Cork-street on one day reached 185, these being nearly all typhus cases. From 1866 (year ending 31st March, 1867) fever steadily decreased until the year 1869, when the admissions reached but 1,823. It has, however, been since rising, the admissions being 2,264 and 2,343 respectively, for the two years ending March 31st, 1871. It will thus be seen that the rate of ad-

missions to the Fever Hospitals was much the same during
the year ending March 31st, 1871, as it was ten years ago.
The total number of admissions were 42,534.

TABLE I.—Showing the ADMISSIONS into the CORK-STREET FEVER
HOSPITAL, and the HARDWICKE FEVER HOSPITAL, from the year
ending March 31st, 1857, to March 31st, 1871.

Year ending	Admissions into Cork-street Hospital.	Admissions into Hardwicke Hospital.	Total Admissions.
(1)	(2)	(3)	(4)
March 31st, 1857, . . .	1,606	1,705	3,311
„ 1858, . . .	1,466	1,626	3,092
„ 1859, . . .	1,310	1,609	2,919
„ 1860, . . .	1,616	1,430	3,046
„ 1861, . . .	1,478	1,174	2,652
„ 1862, . . .	1,700	1,129	2,829
„ 1863, . . .	1,845	1,179	3,024
„ 1864, . . .	1,747	1,405	3,152
„ 1865, . . .	2,086	1,249	3,335
„ 1866, . . .	2,151	1,411	3,562
„ 1867, . . .	1,774	1,379	3,153
„ 1868, . . .	1,098	931	2,029
„ 1869, . . .	965	858	1,823
„ 1870, . . .	1,270	994	2,264
„ 1871, . . .	1,357	986	2,343
Total, . . .	23,469	19,065	42,534
Average, . . .	1,564	1,271	2,835

I have shown what a large proportion of the death-rate is
caused by Zymotic diseases, namely, one-fourth in England
and Scotland, and one-fifth in Ireland, or, more exactly,
21·9 per cent. of the total mortality of the United Kingdom.
The chief portion of these deaths is concentrated in the large
towns, and is the chief cause of the town death-rate being
in excess of the country death-rate. From this we may
conclude that the variations in death-rate in large towns will
follow very closely the variations in prevalence of Zymotic
disease; and the relations in death-rate between different
towns will also closely follow the relative prevalence of
Zymotic disease. This is well demonstrated in the accompanying tables and diagrams, showing the relation between
the variations in death-rate from Zymotics in Dublin, as
compared with that of the total death-rate, and the relation
between the Zymotic death-rate in large towns, to the total
death-rates.

On looking at Diagram I. representing, for the past nine
years, the relations between total deaths in Dublin, total

deaths from Zymotic diseases, deaths from fevers, as a specimen of the result of endemic diseases, and deaths from
cholera in 1866, and small-pox in late epidemic as specimens
of the results of the only two great epidemics which occurred
during this period, it will be evident that, with few exceptions, the variations in the Zymotic deaths correspond very
closely with the variations in total deaths. The chief
exceptions may be easily explained, and have already been
referred to by Dr. Moore ; they depend upon the effect of
cold, especially intense cold, increasing the deaths from chest
diseases to an enormous extent. The best example of this
was the intense cold which followed the subsiding of the
cholera in 1866, and the cold of the beginning of 1871,
specially mentioned by Dr. Moore, for, after the cholera had
subsided, the great cold of January, 1867, raised the death-
rate higher than it had reached during the height of the
cholera epidemic. You see that the curve representing the
total deaths is accompanied by one representing the average
weekly deaths for the corresponding periods, and we see
that, with very few exceptions, where the total death curve
rises above the weekly average, that it corresponds with a
rise in the Zymotic curve, excepting always the increase
owing to great cold, which can easily be corrected by observing the depressions in the curve of weekly mean temperature
added to the diagram by Dr. Moore. Again, if we arrange
the following large towns according to their death-rate,
placing the one with the lowest death-rate at the top of the
list, and in a parallel column arrange the same towns according to the death-rate from Zymotic diseases, we shall find
a close correspondence :—

TOTAL DEATH-RATE.	ZYMOTIC DEATH-RATE.
1 Birmingham,	1 Bristol,
2 Hull,	2 Hull,
3 London, Bristol,	3 Birmingham, Leeds,
4 Dublin,	4 Edinburgh,
5 Sheffield,	5 London,
6 Edinburgh,	6 Dublin,
7 Leeds,	7 Sheffield,
8 Newcastle-on-Tyne,	8 Liverpool,
9 Salford,	9 Manchester and Newcastle
10 Manchester,	10 Glasgow,
11 Liverpool, Glasgow.	11 Salford.

Hull is the only one where this correspondence is exact,
it standing 2nd on both lists. You may also see that although
Bristol stands highest as being most free from Zymotics, yet

it is 3rd in the death-rate list; Birmingham being first on
this list, and 3rd on the Zymotic list. The cases of Birmingham and Bristol I shall refer to again. Dublin, though
tolerably high in the death-rate list, is about average so far
as Zymotics are concerned.

A reference to Table II. and Diagram II., constructed to
show the relations between total death-rate, Zymotic death-
rate, and density of population, in the 13 large towns referred
to, will also show a marked amount of parallelism between
the curves representing the Zymotic and total death-rates;
Bristol being a well marked exception, and this exception is
to be altogether attributed to the extensive measures undertaken to improve the sanitary condition of the city, and to
maintain the public health.

TABLE II.—SHOWING the relation between density of Population,
general Death-rate, and Death-rate from ZYMOTIC DISEASES in 13
large Towns of the United Kingdom, as shown in Diagram II.

Towns.	Population per Acre.	Deaths per 1,000.	Deaths per 1,000 from 7 principal Zymotics.
London,	41·8	24·2	4·5
Bristol,	37·0	24·2	2·0
Birmingham,	48·3	23·4	3·6
Liverpool,	103·0	31·1	5·6
Manchester,	84·5	30·4	6·0
Salford,	23·9	28·5	7·5
Sheffield,	11·2	26·5	4·6
Leeds,	12·3	27·2	3·6
Hull,	38·0	24·1	2·8
Newcastle-on-Tyne,	25·5	28·0	6·0
Edinburgh,	40·6	26·9	4·0
Glasgow,	94·3	31·1	6·4
Dublin,	33·1	25·4	4·4
Total,	45·6	25·8	4·7

In 4 towns the density of population above average.
In 3 of those mortality is above average; in 1 below—Birmingham.
In 3 the Zymotic mortality is above average; in 1 below—Birmingham.
In 9 the density of population is below average.
In 5 of those mortality is above average; in 4 below average.
In 2 of those Zymotic mortality is above average; in 7 below average.

The same proposition that Zymotic death-rate has a nearly
constant relation to total death-rate is also shown by a
comparison of the 28 districts of the London Registration
Division for the year 1870, as shown in Table III. and
Diagram III., constructed on the same principle as that for
the 13 large towns mentioned above.

DIAGRAM L

Showing the relation between Fund Bottles, Total Bottles from Bywater Blowers, and the items Fund, Orders, and Brandigan, weekly for 9 years,
ending December 31st, 1872. Compared with the mean barometer for the same period

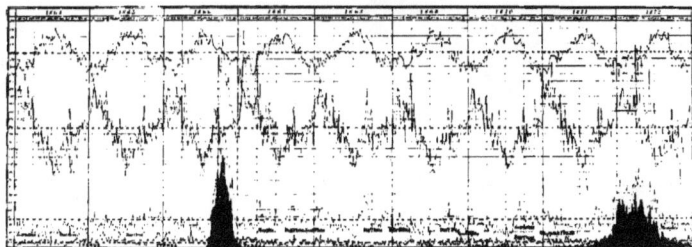

TABLE III.—SHOWING the Relation between Density of Population, General Death-rate, Death-rate from ZYMOTIC DISEASE, and Pauperism, in the London Registration Districts, as shown in Diagram III.

District.	Population per Acre.	Total average Death-rate per 1,000 for 10 years.	Death-rate per 1,000 from 7 Zymotics, 1871.	Paupers per Population in London District receiving Out-door Relief, 1871.
Kensington, . . .	39·3	19	4·5	1 in 36·9
Chelsea, ...	82·2	25	6·6	,, 47·0
St. George's, Hanover-sq., .	74·1	19	4·9	,, 30·2
Westminster, . . .	235·7	23	4·2	,, 40·1
Marylebone, . . .	105·4	24	4·2	,, 41·0
Hampstead, . . .	14·3	17	4·4	,, 122·2
Pancras, ...	81·5	22	4·8	,, 25·5
Islington, ...	68·0	21	7·2	,, 42·2
Hackney, ...	31·7	19	3·8	,, 20·7
St. Giles', ...	217·0	28	4·7	,, 36·7
Strand, ...	97·2	22	4·2	,, 34·5
Holborn, ...	205·2	26	4·1	,, 21·9
London City, . . .	104·6	19	4·0	,, 13·9
Shoreditch, . . .	196·8	26	6·1	,, 30·4
Bethnal Green, . . .	158·2	23	5·0	,, 32·4
Whitechapel, . . .	185·6	28	4·6	,, 22·3
St. George's-in-the-East, .	198·8	29	5·5	,, 11·9
Stepney, ...	99·9	27	4·2	,, 19·3
Mile End, Old Town, .	136·8	24	4·7	,, 29·0
Poplar, ...	39·4	24	4·5	,, 25·6
St. Saviour's, Southwark, .	151·3	29	4·9	,, 22·6
St. Olave's, Southwark, .	69·4	29	4·4	,, 30·9
Lambeth, ...	51·8	23	5·3	,, 32·2
Wandsworth, . . .	10·7	20	5·3	,, 27·4
Camberwell, . . .	25·6	23	3·6	,, 30·6
Greenwich, . . .	26·6	24	5·2	,, 21·7
Lewisham, . . .	4·5	18	4·2	,, 12·6
Woolwich, . . .	9·9	–	6·7	,, 44·4
Average, . .	41·8	24·2	4·4	1 in 27·8

From an analysis of the tables from which these curves are constructed, we find that of the 13 large towns enumerated the death-rate is above average in 8, and of these 8 the Zymotic death-rate is above average in 5, nearly at average in 1, below average in 2, and of those where the death-rate is below average, the Zymotic death-rate is never above average, and nearly up to average in but one case only.

I have thus shown what a large number of lives are lost owing to this class of diseases, and especially owing to those to which I have especially referred; also what an influence Zymotic diseases have on the total death-rate, especially that of large towns. Besides the absolute loss of life—to this loss of life from preventable disease the term " *life waste* " has

been applied by my friend Dr. Maunsell, a high authority
on the economic relations of disease and pauperism—the
money loss is immense to the community besides to individuals.
Thus the small-pox, the most severe epidemic that has visited
Dublin this century, except the great fevers of 1826, and
famine fever of 1847, and cholera epidemic of 1832 and 1849,
cost Dublin not less than £35,000, probably £40,000, for the
treatment of the sick. Nearly the whole of this was spent on
hospital appliances, food, medicine, and stimulants for the
sick, the medical attendance costing a mere trifle, probably
not £1,000 for poor law medical relief. Besides this, there
was great loss to the trade of the city. The records of the
Mansion House Small-pox Relief Committee show the incalculable amount of misery caused by epidemics. These show
that the applicants for relief represented 6,000 persons who
were affected by small-pox, and who were reduced to apply
for relief on account of the loss caused by the disease to
them and their friends ; 667 heads of families suffered, 179
heads of families died.

None but clergymen and medical men know of the wide-
spread anguish caused by this terrible visitation. The clergy
of all denominations were morning, noon, and night passing
from house to house, and bed to bed, comforting the sick
and dying, the widows and orphans. We members of the
medical profession had to visit our hospitals not only once
but twice, and often thrice daily; the dispensary medical
officers spent their whole time in the dens of this noisome
pestilence. As an example of the amount of work thrown
upon us by this terrible but preventable disease, there fell
to my share within a few of 600 cases, and other members
of the profession had to deal with perhaps larger numbers.
With such an example as this almost present before our eyes,
it is unnecessary to look for other illustrations from more
remote or less striking evils ; Dr. Stokes has already described
to you the terrible results of the famine fever of 1847. After
what I have said, and from the important place that small-
pox and cholera occupy in Diagram I., you will perhaps be
surprised to hear that such epidemics are but secondary in
their destructiveness ; they are more sharp and decisive, but
other diseases such as fever, scarlatina, measles, and diarrhœa,
do their work of destruction quite as surely, only creating
alarm where they rise with epidemics, as indicated in the
diagram where their names were introduced between the
zymotic and fever cases. Let us compare the relative
amount of damage done by these various zymotics in Dublin
during the space of nine years, while there has been a

systematic registration of deaths in Ireland. This is shown by diagram IV. and the following table:—

		Total for nine years.	Average rates per annum.
1. Fevers,	. . .	3,506	389
2. Diarrhœa,.	. .	2,576	286
3. Scarlatina,	. .	2,407	267
4. Smallpox, .	. .	1,699	188
5. Whooping-cough,	.	1,464	162
6. Cholera,	. .	1,293	143
7. Measles,	. .	1,124	124

From this you will see that fever is by far the most destructive, and the per-centage of deaths of those attacked being low, the 3,506 deaths represent an enormous number of cases, probably not less than 50,000 severe cases. As fevers are chiefly fatal to adults, while scarlatina, measles, diarrhœa, and whooping-cough are chiefly fatal to children, it is evident that the relative misery and loss produced by fever is greater than any of the other zymotics, and I wish especially to insist on this point that the endemic zymotics constantly among us, any of which may assume the epidemic form, are far more destructive than those which only appear as epidemics; and however much it may be our duty to ward off cholera and small-pox from our shores, yet it is equally our duty, and far more important for our national prosperity and domestic comfort, that we should control these endemic diseases which never die out. The foregoing statements refer not only to Dublin, but to nearly, if not to all, large towns. As I have shown by the diagram how much zymotic death-rate influences the general death-rate, so I can here show how much the fever death-rate influences the zymotic death-rate, and therefore influences the general death-rate more fatally than any single disease, except consumption.

If you look at the curves on diagram I. you will see that the elevations and depressions of the zymotic curve correspond almost invariably with elevations in the fever curve, although zymotics are sometimes high when fever is low, as seen in the diagram, where the various other zymotics which, together with fever, went to make up the total zymotic mortality, are marked between the zymotic and fever lines. There was but one week during the whole nine years represented on diagram I. in which no deaths from fever were registered. There are good reasons why this correspondence should exist, as I shall show when I discuss the conditions under which zymotic diseases spread.

G

CONDITIONS UNDER WHICH ZYMOTIC DISEASES SPREAD.

The conditions which influence the spread of zymotic diseases are numerous, but are easily classified; they belong to one of two great classes—those belonging to places, or those belonging to persons or population.

A. Those belonging to place—

1. Locality, whether high or low, elevation, and geological formation of ground on which it lies.
2. Facilities for drainage and water supply.
3. Age, condition, and construction of streets and houses.
4. Climate.

B. Those which belong to population—

5. Density of population.
6. Proportion of pauperism.
7. Cleanliness of inhabitants.
8. Accommodation for the sick.

Locality.

It is a well established fact that the higher the situation above the level of the sea the less the prevalence of zymotic diseases; this no doubt is chiefly owing to the facilities afforded by such situations for efficient drainage, and also to the fact that few large communities are so situated. As, however, the situation of all our towns and most of our villages has already been settled, we may almost leave this consideration out of the question. Their position can scarcely be at all effected by public measures, but the defects in situation may be counteracted to a great extent by sanitary measures. As an example of the effects of situation, Birmingham, which from its elevated position, porous soil, and favourable aspect, has a system of natural drainage, although densely populated and without any particularly good sanitary system, is the healthiest of the large towns in the United Kingdom, usually escapes great epidemics, and has a low zymotic death-rate.

Drainage and Water Supply.

The great effect of proper drainage and water supply on the health of towns is shown by the instances given in the accompanying table IV., taken from the Ninth Report of the Medical Officer of the Privy Council, from which it appears that a large number of English towns have been materially

improved in the health of their population by extensive improvements in drainage and water supply; endemic diseases have permanently diminished, and epidemics have fallen with lightness on these towns since the improvements have taken place. This is especially and almost invariably to be noticed in enteric fever, diarrhœa, and cholera, the only diseases I considered necessary to include in my tables.

This table shows the result in ten of the largest towns enumerated in the report, several others, however, have undergone similar improvements by similar means.

Age, Condition, and Construction of Houses.

It is a notorious fact that old houses in old streets of old towns are the favourite haunts of zymotic disease, and there are other reasons for this besides the age of the houses, for it is here that we find poverty, hunger, and dirt, combined with overcrowding, all of these being promoters of zymotic disease. A comparatively new house may, too, from faults in original construction, want of drainage, and neglect of repairs and cleaning, become as bad as any old house.

I have constructed from the records of the Cork-street Fever Hospital a list of all the houses on the south side of this city, from which cases of fever were admitted into the ⌐pital during a period of two years ending September 30th, ⌐71. These houses are marked on this map by red dots.* Two lines intersect the map, and the point of intersection marks the centre of old Dublin as it existed in A.D. 1610; the boundary of the city at this date is also marked on the map, and the boundary of the city in 1728. A glance at the map will show that by far the greater number of fever dots are concentrated in the area of the old city, the remainder being nearly contained between the boundaries of 1610 and 1728, the next oldest part, and but few are situated beyond the latter line or modern part of the city. I have traced out 1,190 of these fever houses; of these there were 122 especially productive of fever, and of these 122 no less than 70 are within the old city boundary. The worst fever streets in Dublin are to be found among the oldest, thus, Francis-street which was fully built in 1610, has 28 infected houses out of 140. The Coombe, though not so old, and Meath-street more modern still though old, are remarkably productive of fevers. The same may be said of the old streets lying along the bank of the river, though not on the line of

* An ordnance map with these and other markings mentioned in the lecture was exhibited at the time of its delivery.

TABLE IV.—ILLUSTRATING the Improvements of PUBLIC HEALTH resulting from Proper Works of Drainage and Water Supply.

Population in 1861.	Name of Town and order of Population.	Periods for which comparisons were made.		Deaths per 10,000 for each period compared.				Cholera in each of Three Epidemics.			General Death Rate.	
		Before Works.	After Works.	Enteric Fever.		Diarrhœa.		1848-9.	1854.	1866.	Before.	After.
				Before Works.	After Works.	Before Works.	After Works.					
160,714	Bristol,	1847-50	1862-5	10	6·5	10·5	9·5	82	11	1·5	245·5	242
68,056	Leicester,	1845-51	1862-4	14·6	7·7	16	19·3	1	10	–	264	252
52,778	Merthyr,	1845-55	1862-5	21·3	8·6	11·5	6·2	267	84	20	332	262
39,693	Cheltenham,	1845-57	1860-5	8	4·7	8·3	7	–	–	–	194	185
32,954	Cardiff,	1847-54	1859-66	17·5	10·5	17·2	4·5	208	66	15·5	332	226
30,229	Croydon,	1845-50	1857-64	15	5·5	10	7	27	21	2	237	190
29,417	Carlisle,	1845-53	1858-64	10	9·7	11·3	12·5	22	6	–	284	261
27,475	Macclesfield,	1845-52	1857-64	14·2	8·5	11·3	11	9	1	–	298	237
24,756	Newport,	1845-49	1860-65	16·3	10·3	11	6·5	112	1·5	12	318	216·5
23,108	Dover,	1843-53	1857-65	14	9	9·5	7	40	10	4·7	225·5	209

our modern quays. In old times there was no "north side" of Dublin except a small portion around St. Michan's Church and the Abbey of St. Mary, and following the rule I have laid down, these are the worst fever streets on the north side of the city. Now why have I said so much about the localities where fever prevails? because I have ascertained that these also are the places where *all* zymotic diseases arise and spread; thus a cholera map or a small-pox map would be precisely the same as this fever map; the same streets would be in the same colours, and many of the same houses would be marked in both; thus, of 124 fever nests, 58 at least have been also small-pox and cholera nests. This is not true only of Dublin but also other large towns as was shown in the *Lancet* report on "Cholera Haunts and Fever Dens of London," and Dr. Gairdner's remarks on the fever dens of Glasgow point to the same conclusions; but it is unnecessary to go out of our own city to find proof of this. On again referring to the map it will be seen that a number of circles are drawn round certain localities. These represent especially infected places, which are the constant habitats of fever and diarrhœa, and where I have ascertained that cholera and small-pox have prevailed in the last two epidemics—in some instances even in the epidemic dating as far back as the cholera of 1832. I may here describe by extracts from my paper on the "Prevalence and Distribution of Fever in Dublin," and from the report of the Dublin Sanitary Association, the conditions we find in these fever streets and houses :—

"The streets are generally characterized by being composed of old—many of them once fashionable—houses, with bad rears, or no rears at all. It is not essential, as many suppose, that fever streets should be narrow and tortuous; on the contrary, two of the worst fever streets, Meath-street (the very worst), and Francis-street, are wide and straight. It is the age and condition of houses, and proximity of narrow courts and alleys, that especially characterize these streets, together with the want of proper house drainage, ash-pit, and privy accommodation for the houses themselves. As examples of the worst fever streets, I may mention Meath-street, with its 95 houses, 36, or more than a third, of which furnished in all 73 cases of fever to Cork-street Hospital during the two years ; it contains one fever-nest furnishing 6 cases, and 10 others furnishing 3 or 4 cases each. Francis-street, with 140 houses, has 28 fever houses, furnishing 55 cases, has in it two fever-nests furnishing more than 5 cases, and 6 houses furnishing 3 or 4 cases each. The Coombe contains 129 houses, has 46 fever houses, furnishing 78 cases, one house furnishing 5 cases, and 4 others furnishing 3 or 4 cases each. These are sufficiently detailed examples of fever streets; but I could mention many others nearly, though not quite so bad. The lanes and alleys are probably worse than the streets, but must be merely looked upon as streets on a smaller scale. The courts (comprising yards and squares) are next to be considered. These are, perhaps, the most prolific fever beds, as few of them have failed to produce fever cases during the past two years. Fever streets are generally skirted by these courts, notably those which I have already given as special examples of fever streets. There are several kinds of courts—first, those originally constructed as such ; secondly, lanes closed up at one or both ends, and entered by archways ; and thirdly, back yards and gardens that have, by the cupidity of the owners, been built upon, and the out-offices converted into dwelling-houses, thus crowding together a large number of small tenements in a very confined space. These latter are generally known by the name of yards, and are usually designated by the number of the house behind which they are situated. Few people besides clergymen and medical men are acquainted with the existence of these places.

"Examples of the first form of court may be found in abundance off South Great George's-street and Kevin-street, and a considerable number in the neighbourhood of Townsend-street. They are, in fact, narrow, blind lanes, and have usually an open sewer running down the centre through the whole length, and emptying itself into the adjoining street, or into a trap near the entrance of the court. These traps are frequently choked, and large quantities of sewage accumulate. There is usually a privy, seldom more than one, situated in each of these, as also in the other form of courts ; the drainage from this privy, of course, finds its way down the open sewer already described in the centre of the court. The square may be considered as the last of these forms of court, samples of which are Gill's-square, off Cole-alley, Neil's-court, off Marrowbone-lane, Derby-square, off Nicholas-street, &c. These squares have usually no drainage, and are surrounded by miserable old overcrowded houses, and are generally strewn with rubbish and filth, consisting, to great extent, of human ordure, and have one or two cess-pools near the centre. I have already indicated the nature of the yards, several of which may be found in Marrowbone-lane, Cork-street, and the Coombe. The houses in all these are of the most filthy character, and the front house or houses in the street usually indicate the nature of what is behind, having the usual characters of a fever-nest, which I shall presently refer to more particularly. The ground of all these courts is saturated with decomposing organic matter, chiefly human excrement."

The following is taken from the Report of the Dublin Sanitary Association :—

October 4th, 1872.

"*Nos.* 17 *and* 18, *Great Ship-street,—Overcrowding—fever—* unfit for habitation in their present state.

"*No.* 17—*Basement Story* damp and filthy—sewage matter is said, at times, to ooze up through the flagging.

"In the kitchen   *front* 2 persons live.
„   ground floor *front* 7   „   „   3 children sick.
„   „   „   *back* 4   „   „   —
„   first   „   *front* 8   „   „   1 person sick.
„   „   „   *back* 3   „   „   1 person sick.
„   second „   *front* 5   „   „   —
„   „   „   *back* 5   „   „   —
„   top   „   *front* 5   „   „   —
„   „   „   *back* 1 person lives   —

"Total population consists of *forty* persons, *five* of whom are now sick, suffering chiefly from various forms of fever.

"*No.* 18—*Basement Story* uninhabited, but filthy. The population of the rooms is as follows :—

Ground floor—*front and back*—4 persons.
First   „   *front*   .   .   6   „
„   „   *back*   .   .   5   „
Second   „   *front*   .   .   8   „
„   „   *back*   .   .   9   „
Top   „   *front*   .   .   7   „   1 case of fever.
„   „·   *back*   .   .   4   „

"Total population is, therefore, *forty-three* souls. Total population of both houses, *eighty-three* souls, of whom *six* are now ill.

"In the rere of these two houses three cottages are situated.

"No. 1 is inhabited by a family numbering *five* persons, one of whom is ill of fever in Cork-street Hospital, and a second was to-day removed to the Meath Hospital, suffering from phthisis. (She died October 7th, 1872.)

"No. 2 has also *five* inhabitants—two of whom are now in fever—one in hospital, the other, a young child, at home. This, as the other cottages, is divided into two rooms by a wooden partition, not reaching to the ceiling. The total dimensions are 14 by 8½ by 11 feet, the space per head being only 262 cubic feet.

"No. 3 has *seven* inhabitants, two of whom are now suffering from fever in Cork-street Hospital. The father is in extreme danger, having an attack of severe maculated typhus. The rooms in this cottage are very dirty compared with those in Nos. 1 and 2.

"The total population of the holding 17 and 18, Ship-street, is exactly *one hundred* souls.

"The sanitary accommodation consists of a Vartry-water tap, two privies (each with *two* seats), in average order, and one large ashpit, which requires cleansing.

"Your Sub-Committee would call earnest attention to the formidable outbreak of fever which has taken place in these houses—

due, in a great measure, to the great overcrowding of the rooms, and to the defective sanitary condition of the basement storey of the houses. No less than 11 per cent. *of the population are at present stricken down by fever.*

"These houses are built on the site of an old graveyard; they face the Ship-street Military Barracks, and a line of stables is situated to the south side of them."—*Report of Dublin Sanitary Association.*

"What are the characters of a fever-nest? The best way to answer this question is by describing one or two. I shall begin with the worst on my list, 58, Bridgefoot-street. This house is entered from the street by a passage, with a black and rotten floor, in which are open chinks communicating with the cellar below; the boards are damp and sodden with dirt. Going upwards, we find things somewhat better, but the whole upper part of the house is dilapidated. Going downwards, we first come to the entrance of a small back yard, a place covered ankle-deep with human filth, a privy and ashpit totally unapproachable without passing through a sea of dirt, a water-tap running, and washing such of the dirt as is within reach into a pipe sewer which runs through the cellar of the house, and which has a hole through which the sewage passes into the cellar, converting it into a cesspool; this cellar is immediately beneath two rooms inhabited by a family of fifteen, every one of whom has had enteric fever. In the same street I find another house with all these characteristics repeated, except the broken sewer, but this house had no sewer at all. A house in Chancery-lane furnished eight cases of fever (seven typhus and one enteric). I was met on entry by a horrible stench, proceeding partly from a filthy back yard, and partly from a slaughter-house at the rere of a neighbouring house in Bride-street. The cellar of this house had been filled up—a very proper measure, if rightly carried out; but the filling-up matters consisted of such material as to convert the cellar into a decomposing manure heap. The passage, back yard, and upper part of the house were similar to those already described at 58, Bridgefoot-street. I find similar conditions, varying only in degree, in almost every fever-nest. The less prolific fever-nests I find with less accumulated dirt, and notably less wet dirt. In many places where there was comparatively little dirt, what did exist was made to do the maximum amount of damage by being kept in a continual state of moisture, for want of proper drainage, or from drainage water from the roof or elsewhere running into the house by the doors, or through imperfectly closed cellar openings. These damp cellars, often nearly filled with rubbish, are to be found in all fever streets and most fever houses. Many houses have no receptacle for rubbish except the cellars; this is particularly true of corner-houses and houses near corners, many of which, if not public-houses, are fever-nests. Of the condition of these houses I may also state that a large

DIAGRAM II.

Showing the relation between Densinity of Population,
Total Death Rate, and Death Rate from 7 Principal
Zymotics in 13 large towns of the United Kingdom.

The Shading represents the Density of Population.

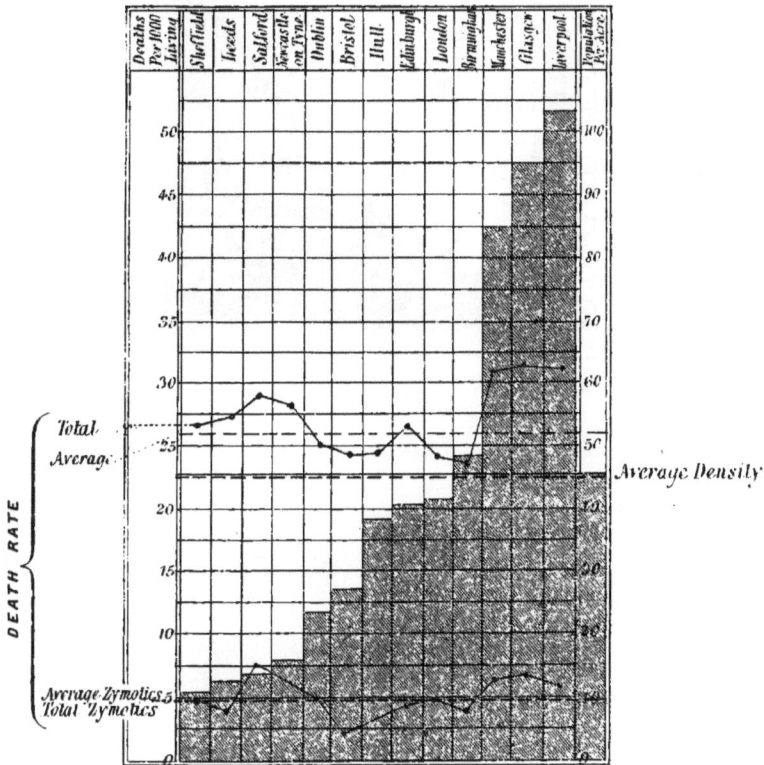

DIAGRAM III.

Showing the relation between Death Rate from all causes, Death Rate from principal Zymotic Diseases,
Pauperism and density of Population in the London Registration District.

The Shading represents the Density of Population

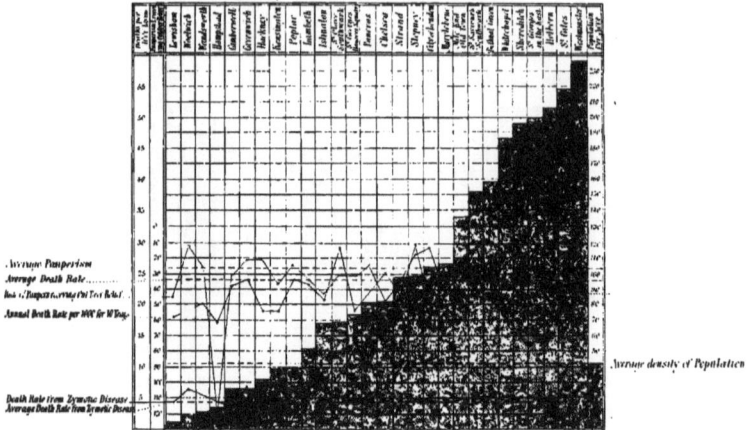

Average Pauperism
Average Death Rate
Rate of Pauperism arising Out Test Relief
Annual Death Rate per 1000 for 10 Years

Death Rate from Zymotic Disease
Average Death Rate from Zymotic Disease

Average density of Population

number are condemned houses—that is, houses declared by the authorities as unfit for human habitation ; but, through some evasion, or, in many cases, in open violation of the law, these houses are still inhabited, and frequently the occupiers even pay rent. I find that among my list of fever houses, there are fifty-two returned as bankrupt in the last report of the Collector-General of Rates ; thus, the owners of these (who defy the tax-gatherer) and of the condemned houses, go free of the burdens of ordinary citizens, and claim as their privilege to spread disease and death at the expense of their honest, and, perhaps, not more prosperous neighbours."—*Prevalence and Distribution of Fever in Dublin.*

We next come to the conditions affecting the population themselves, as promoting causes of Zymotic diseases ; and here we meet with great difficulties in the measurement of each of the various factors of density, poverty, and dirt of the population. We might expect that one—namely, poverty—would regulate the other two, but it does not altogether do so, although it is one of the chief regulators.

Density of Population.

Density of population seems to be the great promoting cause of Zymotic disease. This is shown by a comparison between the various London districts and between 13 large towns of the United Kingdom, as shown in diagrams II. and III. and tables II. and III. Of the 13 large towns—in 4 the density of population, as measured per acre, is above average ; in 3 of these the death-rate is above average, and in these 3 the Zymotic death-rate is above average, the four being Birmingham, which is an exception to all such rules. Of the 28 London districts—in 20 the density of population is above average ; in only 12 of these is the total death-rate above average, but in 17 out of the 20 (or 85 per cent.) the Zymotic death-rate is above average. This, I think, proves the connexion between density of population and Zymotic death-rate.

Pauperism.

That pauperism has also a considerable influence in promoting Zymotic diseases, may also be shown from the London districts ; for in 12 of the 28 in which the pauperism was above average in 9 (or 75 per cent.) the Zymotic death-rate was above average—not so great as density of population, which gives 85 per cent. It is important here to show that the density of population and pauperism do not correspond ; for in the 20 districts in which the density of population is above average, in only 8 is pauperism above average, and in 12 it is below. As might be expected, in these 8 both pauperism and density of population increased the Zymotic mortality. Therefore, though density of population is not the exact out-growth of pauperism, yet where both are combined the Zymotics prevail to the greatest extent.

Cleanliness.

The next point is the effect of dirt in the promotion of Zymotic diseases ; and that dirt has a large share in its propagation I have no doubt.

We cannot as yet certainly ascribe any particular kind of disease to any particular kind of dirt. The characters which I have described as pertaining to fever streets and fever houses are pretty well known to sanitarians, and these are the resorts of all kinds of Zymotic diseases, where dirt is the most prominent character. Now, what is dirt in considering it in connexion with disease ? It will not do merely to say that " dirt is matter out of place "—an excellent definition, nevertheless. I do not mean mere personal dirt, but dirty air, dirty food, dirty drink, dirty houses, dirty clothes, and dirty persons, all combined ; and these kinds of dirt mean *poison* for human beings, and manure for Zymotic diseases to flourish upon. The refuse of cowsheds and stables is dirt in a farm-yard, but it is food for the crops, and ends in being food for ourselves when, in nature's laboratory, it has been converted into bread and beef. Thus our domestic refuse is dirt when in and about our houses, or when it gets into our food or drink, or contaminates the air we breathe, but on the farm may be useful, for in the fields it is manure for the crops, but in the house is manure for the fertile growth of all Zymotic diseases.

I shall show you presently how dirt in air, water, and food, may and probably does cause, certainly promotes, special kinds of disease. Cleanliness, as contradistinguished from dirt, means the removal of all useless or injurious matters from our persons, dwellings, food, and drink, the principal means adopted for which purposes are washings and scrubbings; and if we admit—which I think few will have the hardihood to deny—that some diseases are promoted by contagion, it is manifest that in cleaning away dirt, we are removing the dirtiest though most minute of all dirt, the contagious particles of disease.

Accommodation for the Sick.

Another promoting cause of Zymotic disease is the want of proper accommodation for the sick, namely, proper and easily expansible hospital accommodation for all forms of contagious diseases, the want of proper means of bringing patients thither, the want of proper means of disinfection and separation of the sick and convalescents from contagious Zymotics from the healthy. I regret to say that all these wants exist to a lamentable extent in this city.

In pointing out what are principal promoting causes of all Zymotic diseases, I think I have shown pretty clearly how epidemic Zymotic diseases attack favourite haunts time after time, and that these same haunts are the favourite habitats for fever and diarrhœa at all times. The President of the College of Physicians in the next lecture will, I believe, show you conclusively some of the ways in which these conditions make those subject to them liable to disease.

I now proceed to point out the origin of some of the more important of these Zymotics, or, more properly, to show what conditions seem absolutely essential to the production of them. There are many Zymotics which never seem to arise without contagion, although they flourish best on the soil prepared for them in the ways already described. It is not because we cannot always trace contagion that we are to deny its existence, as I think will be clearly proved by Professor Haughton. As the subject of contagion will be fully treated of by that eloquent Professor, it is quite unnecessary for me to refer further to it, but merely state that evidence of the introduction of the most alarming epidemics of cholera and small-pox through contagion is overwhelming, and it is scarcely necessary to say that overcrowding and dirt promote contagion. I have now to consider certain conditions which have appeared so constantly in connexion with outbreaks of certain epidemics, that they seem to be essential to the production of these diseases, and may therefore be put down as causes of disease. The best established of these are the production of cholera, diarrhœa and enteric fever, in connexion with water contaminated with sewage matter, and air polluted with sewage gas; also the causation of typhus fever by overcrowding. Besides these, we have it suggested, and substantial evidence given in support of the suggestions, that measles is produced by a miasma arising from decomposed vegetable matter, and that scarlatina is produced by decomposed blood and slaughter-house refuse.

But now, to consider these separately—

1st. *Fevers.*

Of these we have four kinds—simple, typhus, enteric, *i.e.* typhoid, and relapsing fever. The first appears to me to be a mere attempt at one of the other two next, namely, typhus and enteric fever, but not sufficiently developed to be able to identify it with either.   This being my opinion, I believe it may arise under the same circumstances, and accordingly I find that it nearly always prevails where there is typhus or enteric fever, but especially where typhus prevails, as appears from the following table of 42 houses furnishing more than five cases each in two years :—

13 houses furnished cases of three kinds of fever.
19   „   „   „   simple and typhus.
4   „   „   „   simple and enteric.
4   „   „   „   typhus and enteric.
2   „   „   „   simple only.
0   „   „   „   typhus only.
0   „   „   „   enteric only.

Nineteen houses furnishing typhus, furnished also simple fever. The condition which seems necessary to the production of typhus is overcrowding, and although overcrowding favours the spread of all kinds of contagion, yet in no disease has it been so frequently and closely associated with the first appearance of disease as in typhus.   Dr. J. Heysham (1781) traced an outbreak in Carlisle to a house inhabited by 6 families, and where no windows that could be built up were left open, consequently there was no ventilation whatever. Typhus began here without any trace of contagion, and then spread through the rest of the town.   In 1859, typhus fever, which for some months had disappeared from Edinburgh, arose in a poor locality where the houses were overcrowded, and in no instance was there any suspicion of contagion.

"Important evidence in favour of the view that the typhus fever poison may be generated from overcrowding, may be derived from the various records of what have been termed 'Black Assizes,' where Judges, juries, and others in court contracted fever from the exhalations from the prisoners, who, in the days prior to the time of Howard, frequently suffered fearfully from what was termed 'gaol fever' in those days, but which was nothing more nor less than typhus produced by the overcrowding of gaols.   The latest, therefore, most reliable account of a Black Assizes is that of 1750 at the Old Bailey, where 100 prisoners were tried.   These were either placed at the bar or confined in two small rooms opening into the court.   Many present wore affected with a noisome smell. Within a week or 10 days, many of those present were seized with typhus.   More than 40 persons died, including the Lord Mayor, two of the Judges, an Alderman, a Sub-Sheriff, and several of the jury.   Neither the prisoners on trial nor any of those in gaol were affected by fever."—*Dr. Murchison.*

From this account it appears that the poison produced in those prisoners by overcrowding was exhaled by them, and affected those in their vicinity without any other apparent cause.

The following is a good example of the conditions under which typhus fever is produced, mentioned in Dr. Murchison's splendid work on fevers :—

"A court, 11 feet wide, with all matters as to drainage and water supply in good order, and recently constructed. Fever arose in house No. 10, which consisted of 2 floors connected by a narrow staircase.

Ground floors : Cubic Feet.
Front room, . 595 Occupied by a mother and 6 children
Back do., . 544 and grandmother, who came to
nurse those who were sick.
Upper floors :
Front room, . 680
Back do., . 497

"Before the arrival of the grandmother each had 163 cubic feet, after her arrival 142 cubic feet. Windows had all been shut up for the winter, and there was no means of ventilation. The rooms had the well-known animal odour of overcrowded houses."

This is only one sample of the conditions under which typhus is produced (street 36 A). I could mention a number of others from various authors, as also from the reports from the Dublin Sanitary Association and from my own experience. So many instances of this kind being recorded, and so many having been found to be the first beginnings of typhus epidemics, the conclusion is almost irresistible that overcrowding is sufficient to generate typhus *de novo.* An additional, though not absolute, proof that typhus fever is caused by overcrowding is that at the time of year at which overcrowding is greatest, namely, in the winter months, typhus fever prevails to the greatest extent, as already pointed out by Dr. Moore, although, as I have shown elsewhere, during its prevalence it is temporarily increased by increased moisture and temperature.

Enteric Fever.

The evidence that enteric fever is the direct product of food, drink, or air contaminated by the presence of decomposing sewage matter, or by the miasma exhaled thereby, is, if possible, stronger than the evidence of the production of typhus by overcrowding; because in the latter cases many other of the causes favouring the spread of Zymotic diseases co-exist, but in many cases, with regard to the former, nearly all other causes except the presence of the decomposing matter can be excluded. I need scarcely refer, in this assembly, to the sickness of the illustrious Vice-Patron of this Society, His Royal Highness the Prince of Wales, and his hospitable and noble friends, who were struck down with enteric fever at a time when almost, if not quite every cause was excluded except sewage exhalations.

I had recently experience of an epidemic of enteric fever breaking out in a large educational establishment in this city, owing to the want of proper drainage, which led to the contamination of the atmosphere with sewage exhalationsand consequent sickness of a large number of those exposed to its influence.

The class-rooms in which the various students attended lectures were situated in one building. In this same building some of the male students resided. The remainder of the male students resided in a separate house in another street. The female students resided in a third building at a considerable distance from the building containing the lecture-rooms.

In the area of the building containing the lecture-rooms were situated latrines for the use of the male students only; there was also a pump in this area, but at some distance from the latrines, from which drinking water was occasionally obtained. This pump was partly supplied by well and partly by Vartry water, but there was no evidence that this pump supplied the usual drinking water for the male pupils. The situation of the latrines was so low (but slightly above high water mark) that the drainage of these was driven back at each rise of the tide, and when the tide rose very high the area itself was flooded with sewage matter of the most disgusting character. Thus all the gases from the decomposing sewage were mixed with the atmosphere breathed by the unfortunate pupils each time they visited the area. The result was that many of those who frequently visited this area (14 out of about 70), were attacked with fever, and of these 4 died. None of the female students and none of the others who frequented the lecture-rooms and who did not visit the latrines, were affected with fever.

As instances of localized outbreaks of fever depending upon sewage contamination of water, I may mention the instances of Terling, Guildford, and Winterton, mentioned in the reports of the Medical Officer of Privy Council.

Terling is a village in Essex, with a population of about 900 souls, is very much isolated and cut off from communication with neighbouring localities, so much so that most of the inhabitants are related to one another by marriage; their physical and moral characters are both very low. The inhabitants are nearly all farm labourers, living in houses constructed, with few exceptions, of lath-and-plaster, or worm-eaten wood. The people are all badly fed. The cottages are surrounded with almost every conceivable nuisance. "Slops and ashes," says Dr. Thorne, the Inspector who writes the Report, "thrown down in unpaved yards and gardens, manure heaps, cesspools, and masses of decaying vegetable matter"—all rubbish and excreta—lay scattered about in all directions.

"Surrounding one cottage and within a radius of 20 feet, I found one pig-stye, three manure heaps, two cesspools, and a privy the contents of which extended for 12 feet down an adjoining field."

In the centre part of the village, as shown in the map accompanying the Report, each cottage or each group of cottages has its own well, and if the ground is at all undulating this is sure to be situated at the lowest point. All are sunk in the gravelly stratum (which underlies the village) and as a rule uncovered, lined with loose bricks (without mortar or cement), depth 5 to 40 feet, according to height of ground. On a higher level and surrounding these wells, are all the nuisances mentioned above, the drainage from which, owing to the porous nature and lie of the ground, as a matter of necessity finds its way into the wells. None of the outlying houses have wells, but derive their water supply from pools in the fields frequented by cattle and described as "nothing better than stinking pools." Overcrowding was frequent, the sick sometimes being two or three in a bed; in some places 82 cubic feet of space was allowed to each person.

In this village 208 cases of fever of a very bad type occurred besides diarrhœa, and 10 cases of fever in Terling-place, the neighbouring residence of Lord Rayleigh.

The first case of enteric fever arose in the person of Lord Rayleigh's dairy-maid, who drank the water taken from the river Lea in the immediate vicinity of the entrance of a sewer. This dairy-maid had been in Somersetshire but had returned three weeks before she got sick, and as the cases that followed next had no connexion with this case, the supposition of the introduction of the fever by this person is unlikely; it is more than probable she had also opportunities of using water from some of the village wells.

The epidemic followed the rising of the wells after their having been lowered by long-continued dry weather, the rising of the water being caused by wet weather, which, while it filled the wells with rainwater, also washed all the dirt in the vicinity into them.

The first cases which arose after that of the dairy-maid, were in five cottages built of wood, surrounded by pigsties and dirt, all dirt being thrown out into an unpaved yard sodden with dirt, in which yard is situated the well for all the inhabitants of the row.

At No. 1 there was 1 case and 0 death.
,,   2   ,,   1   ,,   1   ,,
,,   3   ,,   2   ,,   1   ,,
,,   4   ,,   3   ,,   0   ,,
,,   5   ,,   2   ,,   0   ,,

The well had dried up and not been used for two months, and water was obtained from a well in the neighbourhood where no great diminution of the water had taken place.

On November 19, a woman was ill, not of the fever, in one of these five houses, and water was wanted for cleansing purposes but none was in the well. On November 26, water was found to be in the well and was immediately used for drinking purposes. Ten days after using this water, or at the end of the time usually allowed for the incubative stage of enteric fever, the first case of fever arose in the person who had used the water, the other cases in these five houses immediately followed the first case.

The inhabitants of another set of houses in the immediate vicinity, who were deprived of their own water supply by the drought, got water from the same well as the first set of houses derived their supply from before the 26th of November. So long as they were thus supplied no fever appeared, but 14 days after the water returned to their own well, fever appeared amongst them in due course of time. The epidemic spread in the same way to the rest of the village, and Lord Rayleigh's house became infected in a similar manner, those being affected only who used water from a well contaminated by sewage matter by leakage from a tank in the neighbourhood of the pump. From the peculiar construction of the house the portion of the inhabitants supplied with water from this source were completely isolated from those in the rest of the house.

Of course all the water in this village must have been contaminated with sewage matter for years. Why then did fever not arise before? The answer is simple; the wells were never before emptied by drought and quickly filled by wet weather, all the dirt being thereby concentrated in the first washing of the sewage-sodden earth.

An almost similar, though not so fatal, an epidemic arose at the village of Winterton. I may give an example of the commencement of this epidemic without further following its details.

A row of four houses were supplied with water from a pump well, within 14 feet of which were situated one open drain, one open ashpit, two pigsties, three privies, and one cesspool, all from 18 inches to 3 feet on a higher level than the well.

In No. 1 there lived 3 persons, 2 of whom had fever.

2	,,	4	,,	4	,,	,,	and 1 died.
3	,,	7	,,	7	,,	,,	
4	,,	4	,,	0	,,	,,	

The people in No. 4 would not drink the water because it had a bad taste, and therefore escaped the fever.

Guildford affords another example of how much influence dirty water has in producing enteric fever. This is a town which, in the year 1861 had a population of 9,000 inhabitants living in 1,675 houses. The town stands on chalk on the side of a hill. The stratum of chalk afforded a natural drainage for the town, all sewage being conducted into cesspools, which drained themselves into the chalk, and therefore remained nearly always dry, and were never known to be offensive. There was no system of drainage at Guildford at time of outbreak. The water supply was derived from several sources—

1st. From an old well sunk in chalk at the bottom of the hill, from which water is pumped by a water-mill.

2nd. A new well from which water was pumped to the upper parts of the town by engine power.

3rd. From private wells attached to the houses.

Nine hundred and twenty-eight houses are supplied from the first two sources, 747 from the private wells.

Some enteric fever always is present in Guildford in its poorer parts.

The outbreak occurred in the last days of August in the upper part of the town where it had not previously prevailed, and where the wealthier part of the population reside. What were the influences to which the infected districts became exposed, which were followed by the fever? There were in all 264 cases of fever, of these 177 were in the 330 houses supplied by the high service water supply, 30 in 598 houses on the low service water supply, and 57 in 747 which received no water from the public water works. This shows at once that those receiving the high level supply were more liable

H

to enteric fever than any others; and excluding the ordinary cases, and the case of children who attended school in the infected districts, but who resided elsewhere, nearly all the fever arose among those drinking the high service water which was drawn from a reservoir in the high part of the town, kept filled from the well by the pumping engine on the low ground.

Up to August 1st these people had a constant supply of water from the high level reservoir; on that day the engine broke down, and the water supply from the old well and wheel was resumed. At this time there was still some water left in the high level reservoir, which was left there exposed to the influences of the heat of August weather, which would promote decomposition of any organic matter it might contain. On the 17th of August the water-wheel broke down, and thus the second supply to the occupiers of the high level district failed. Half a loaf being better than no bread, the small residue of dirty water was supplied to those houses, and then followed the fever. The well from which the water had been pumped was found to be contaminated with sewage, which was comparatively harmless until concentrated and acted upon by exposure to the summer heat while it lay undisturbed for nearly three weeks in the reservoir.

Two points are here illustrated:—1st, how sewage contamination poisons water; and 2nd, how this poison is increased in intensity by decomposition and concentration.

An outbreak of enteric fever at Islington has been shown by Dr. Ballard to have depended upon the sewage contamination of milk from a dairy-yard pump, which had been used to increase the value of the milk to the dairyman, at the expense of the health and lives of his customers.

Of 2,000 families resident within a quarter mile radius of the dairy-yard, 142 were supplied with milk from this dairy, of these 70 were invaded by enteric fever within 10 weeks.

"It is remarkable," says Dr. Ballard, "how typhoid picked out the customers of this dairy; thus in one long road and a street issuing from it, at a distance of a mile or more from the dairy, it supplied 3 families—of these, two had typhoid. It supplied 4 families in a neighbourhood of about 70 houses—of these, 3 had typhoid; it supplied 4 families in a row of 9 houses, typhoid occurred in 2 of them; and in the other 2, cases of a mild febrile character occurred."

And so on in many other instances. Dr. Ballard also shows that only those who consumed the milk were affected by fever, and of those who worked in the dairy-yard and did not use either the dirty milk or dirty water none had the fever. On examination, Dr. Ballard found that the pump from which the milk was, I believe, undoubtedly watered, was contaminated by sewage infiltration into the tank from which the pump water was derived.

Within the last few weeks similar examples have been presented by Dr. Russell, who showed that of 72 families in 5 streets supplied by a dairyman in whose family enteric fever prevailed, 22 had fever; and in 32 families supplied by this dairyman there arose 36 cases of fever.

In conclusion Dr. Russell remarks:

" I regard this as an extreme illustration of what most frequently happens where the sale of articles of food is conducted in close connexion with families, and all their attendant ailments. Milk is, from its composition, a peculiarly favourable medium for the propagation of the germs of disease, and particularly of enteric fever, and it is very likely that many apparently inexplicable outbreaks of enteric fever in families are caused by milk, or even solid food contaminated in the retail shops, especially among the poor. It is a very common practice in all parts of the city for parties to live and rear families in rooms behind shops, through which often the sole access lies, and in which groceries, milk, provisions of all kinds, sweetmeats, fruit, &c., are sold. These shops are 'served' by one or both parents, or by some grown-up child, and when infectious disease enters such a family, it cannot fail to be the source of quite peculiar risk to the public. I have been so much impressed with this by a series of cases in point, that I applied to Mr. Lang, the Procurator Fiscal, to ascertain what legal powers existed to deal with them. Mr. Lang writes his opinion that persons situated as described in the various instances given in your letter have not proper lodging or accommodation. It will, therefore, be possible by this and other provisions of the Public Health Act to deal with such cases, so as to save the poorer classes from the obvious dangers of contagious sickness in such circumstances. I have, therefore, issued to the Sanitary Inspectors an instruction that systematic attention be paid to the health of all families living in the circumstances described, by a more routine visitation than from the character of the people and the locality might be thought necessary. Any case of infectious disease discovered must be specially and immediately reported to the medical officer. The greatest care is to be taken not to injure the interests of the parties referred to by unnecessary publicity in the discharge of this duty ; but at the same time there is a very obvious danger to the public from their private sickness, arising from their mode of living, which quite warrants the interference of the department.'

"The fatal activity of milk as a cause of disease has also been most carefully and scientifically investigated by Dr. Taylor, of Penrith ; Dr. Bell, and Dr. Thorne. It has been shown that not only

H 2

typhoid but small-pox, scarlatina, and even cholera, have probably been communicated to people through the medium of milk. It is, therefore, of the utmost public importance to inquire into the sanitary condition of the cow-sheds and dairy-yards.

"The following is a graphic description of the dairy-yards in the south side of the city, by Mr. Benson Baker, of London, who published some notes on a sanitary tour through Dublin about two years ago. Any one who will take the trouble to investigate the matter now will find it equally applicable :—' In the most densely-populated and fever-infected district, in close vicinity to the Corporation manure depot in Marrowbone-lane, are to be found the cow-sheds and dairy-yards of Dublin. These yards, like the neighbourhood, are abominably filthy ; manure is allowed to accumulate in heaps, from which may be seen small black fetid streams flowing into the open streets. The effluvium from these yards is absolutely poisonous, and is only equalled by the atmosphere in the cow-sheds. In this district man and beast alike fall easy victims to preventable disease.' Speaking of the condition of the cows, he adds—' Dr. Cameron says that the loss from pleuro-pneumonia sustained by Dublin dairymen is at least 10 per cent., yet the dairymen cannot be convinced that the disease is contagious, and, therefore, unless under compulsion from the sanitary authorities, they never disinfect their premises after the removal of diseased beasts from them.' The vital powers of the cows are lowered by their constant respiration in close fetid stables. In some of the sheds the cubic space allowed for a large cow is less than the minimum—viz., 300 cubic feet of breathing room—allowed a man in a registered lodging-house. The cows were so close to each other that it was impossible that they could all lie down together. On questioning the owner on this point he facetiously replied, ' Gorrah, sir, they take it turn about.' This repartee might excite a laugh if the occasion of it did not inflict cruelty on the beasts, and tend to affect the people with disease. It is not surprising to learn that milk obtained from cows herded together in such unsanitary conditions, not only conveys foot and mouth disease, but typhoid and other zymotic diseases to the consumer."

I have little doubt that many cases of fever and diarrhœa are produced in a similar way in Dublin, by the sewage contamination of milk. Any one who visits a Dublin dairy-yard, must have been convinced that the milk derived therefrom must run great chances of sewage contamination of some sort, for more fearfully filthy places can scarcely be imagined. Where causes of enteric fever and typhus co-exist, both diseases will arise at same time and sometimes even in same person, as I have shown to be the case in an account of fever at 50 Bishop-street.—*Irish Hospital Gazette.*

While on this subject I may mention that not only enteric fever, but cholera, small-pox, and scarlatina, are liable to spread in this way. I have treated dairy-maids for small-pox and scarlatina; and I regret to find on Dr. Mapother's street list of cholera in Dublin, in 1866, which he has kindly lent me for the purpose of this lecture, that many dairies are included as having been invaded by this disease. Dr. Taylor has demonstrated how scarlatina was spread in Penrith by means of milk, and similar observations have been made by Drs. Bell and Thorne. I believe I narrowly escaped a visitation of cholera in my own house from a similar cause in 1866. It is to be hoped that if, notwithstanding Dr. Cameron's efforts, our dairymen still persist in diluting our milk they will confine themselves to Vartry water. Dr. Reynolds has shown you how to distinguish in many ways good milk from bad, but unfortunately no means is as yet known for distinguishing milk poisoned by disease germs. The same conditions which favour the spread of enteric fever, also favour the spread of diarrhœa, and in point of fact many deaths of enteric fever, especially in children, are registered as cases of diarrhœa. Enteric fever is a disease of summer, when the decomposition of sewage matter is favoured by the high temperature. Additional proof of the constant influence of poisoning by sewage matter is drawn from the Reports of the Medical Officer of the Local Government Board (formerly of the Privy Council) of England, where in every instance where a town is reported as infected by enteric fever, we find that the arrangements were such that the inhabitants were poisoned by their own sewage.

Relapsing fever is generally believed to be the direct product of famine, but being contagious may communicate itself to well-fed persons. It is as Dr. Stokes informed you, when speaking of the great famine fever, frequently followed by typhus.

Cholera.

In close relation to enteric fever as to causation, stands cholera. The most constant condition connected with the spread of cholera is an impure water supply, or a supply contaminated with sewage matter. I could give numerous instances of this, but shall confine myself to the one of London, where terrible experiments have been carried out on a most gigantic scale, which prove the relation between impure water supply and cholera. I do not mean to say, positively, dirty water produces cholera, but it certainly promotes it, whether by containing the germs of that dire disease, or by merely affording a suitable and apparently necessary soil for the disease to grow upon.

This I may say has been demonstrated by the various effects produced in London by the different cholera epidemics of 1854, 1849, and 1866, on each district according to the nature of the drinking water supplied to the inhabitants. This was first pointed out by Mr. Simon, in his "Report on the cholera epidemics of London, as affected by the consumption of impure water," published in 1856. This Report was the result of most painstaking and lengthy inquiry into the most minute details of water supply, population, and distribution of cholera in 1849 and 1854, in the London districts lying south of the Thames. A similar report by Mr. Radcliffe has been published in the Report of the Medical Officer of Privy Council for 1866, showing the connexion between the diffusion of cholera and impure water supply in the east end of London, in the last cholera epidemic, that of 1866.

In the epidemics of 1849 and 1854, cholera fell with the greatest severity on the portion of London lying south of the river, under the following circumstances:—

There were, and I believe are still, two companies supplying this district (which comprises St. Saviour's, St. Olave's, and St. George's, Southwark, Bermondsy, Newington, Lambeth, Wandsworth, Camberwell, and Rotherhithe) with water, the competition was great between these two companies, so great that out of 31 sub-districts there were but eight which had but one company's mains within it, and in many cases the mains of both companies run parallel in the same streets, supplying about equal number of the houses. Thus the population supplied by the two companies were so intimately mixed, that with the exception of the water supply the conditions were identical. We have thus a most perfect arrangement for testing the influence of bad water in promoting cholera. The two companies in question were the Lambeth Company and the Southwark and Vauxhall Company, supplying a population of about 466,000 in 1849, and about 511,000 in 1854.

In 1853 and '54, the Lambeth Company, which derived its supply from the Thames at Ditton, a source pure (dirty though it may be) in comparison with that of the sister company, supplied 24,854 houses, comprising a population of 166,906 persons, and there occurred 611 cholera deaths, being at the rate of 37 to 10,000 persons living.

The Southwark and Vauxhall Company derived their supply from the Thames at Battersea, which was "found to be of almost incredible foulness," swarming with living things and filled with particles of dirt. In 39,726 houses, comprising 268,171 persons, there occurred 3,476 cholera deaths, or at the rate of 130 to every 10,000 of those living, or about three and a half times as many as those drinking the better water.

In 1854 the Lambeth Company gave the best water, but in 1849 it gave worse than the Vauxhall Company, for the Lambeth Company during the interval moved their works up the river, while the Vauxhall Company remained where they were, and even this source became more impure from the increased drainage poured into the Thames by the increased population of London. Accordingly we find that in the epidemic of 1849, in the houses of the Lambeth Water Company's tenantry, there died no less than 1,925 persons, although the population was less than in 1854 when but 611 died of cholera.

In 1849 there died among the Vauxhall Company's tenantry 2,880, or less than the 3,476 of 1854; making all allowances for increased population, the mortality was higher than 1849, and the water worse. It is thus clear that, in the southern districts of London where the water supply improved, cholera was less, and where it became worse cholera was more prevalent. In 1866, when, by the enforcement of a new Act of Parliament, the Vauxhall Company had been compelled to obtain a new supply, and the Lambeth Company had improved its supply, there was but little cholera on the south side of the Thames. On the other hand, a dirty water supply poisoned the greater portion of the east end of London on the north side of the Thames, as shown in Mr. Radcliffe's report, previously referred to.

The East London Water Company supplied two districts, both of these were infected by cholera, one severely, the other but slightly. There were two sets of reservoirs—one at a place called Lea Bridge, the other Old Ford. The district supplied from Lea Bridge was severely affected; that supplied from Old Ford was terribly swept by the epidemic. But why was this when the water was from the same sources in both cases, and why did not cholera always pervade the population supplied from the Old Ford reservoir? The "Old Ford" reservoir was contaminated by sewage from the River Lea, which at that point was a sort of canal, into which drains emptied themselves, and which were possibly even contaminated by the drainage from the first cholera cases. It was not until in consequence of a short supply of water that this reservoir was used that cholera spread through the district. A map accompanying the report shows, by shadings, the various degrees in which cholera invaded the different districts of London in 1866, and graphically demonstrates how fatally the district supplied by the Old Ford reservoirs of the East London Water Company were affected.

The story of the Broad-street Pump by Dr. Snow further proves the influence of dirty water in spreading cholera, as also did a special outbreak in connexion with a pump in Duke-street, in this city. Thanks to the exertions of my friend and fellow-citizen, Sir John Gray, we are not likely ever to suffer from the effects of an impure water supply; and I have no doubt that when cholera again visits us we shall have few such stories as that of the "Duke-street Pump." But we must not here forget Dr. Reynolds' remarks about the poisoning of water by dirty cisterns.

Valuable evidence in support of the connexion between cholera and water supply is given in Dr. Pettenkofer's papers on the connexion between cholera and ground water; as also the instructing paper by Dr. Mapother on the relation between old rivers and sewers, and the distribution of cholera in Dublin, in which he showed the predilection of cholera for these sites.

Measles.

Next, we have to see under what circumstances measles arise. The only conditions yet shown to be intimately connected with measles are decomposition of vegetable matter, especially straw, and the presence of the lowest forms of fungi, commonly called mustiness.

Dr. Salisbury, of Newark, Ohio, United States, has demonstrated, beyond doubt, his ability to produce measles (or a disease undistinguishable from it), just as the gardener can produce mushrooms by preparing a bed upon which they are to grow.

Dr. Salisbury refers to the various fungi which attack grain as smut and bunt, to those attacking animals as Mursadine (*Botrytis Bassiana*) attacking the silkworm, and the mould which kills the house fly in autumn (*Sporendonema muscæ*), and which we see as white rings around the poor little animal's body, and finally against our window panes after the death of the fly. Many skin diseases are now known to be associated with the production of vegetable growths on the surface of our bodies.

1st. Dr. Salisbury points to the case of Mr. Dill, who got an attack undistinguishable from measles while engaged in turning over a stack of musty straw, the odour from which persistently remained in his nostrils for long after he had done handling the straw.

2nd. In an outbreak of measles at the military camp

near Newark, Ohio, there was no trace of contagion ; the outbreak followed immediately on the melting of the snow while wet, which made musty the warm straw which the men slept upon in their tents.

3rd. Cases mentioned by Mr. S——, in the persons of those employed in thrashing wheat that had become heated.

These cases suggested to Dr. Salisbury the inquiry, whether camp measles were caused by musty straw. He examined the musty straw (wheat straw) to which had been attributed the cause of the measles, he found certain fungi which are figured in his work. He, to prove their identity with wheat straw fungi, grew them in a box.

He then grew some fungi with which he inoculated himself, and produced the symptoms of measles with a partially developed rash ; a second inoculation failed to produce similar effects. Similar effects were produced by inoculation of his wife.

In a family where measles broke out, inoculation by the straw fungi, while giving measles of a modified form, prevented the occurrence of unmodified measles. These are substantiated by other evidence, and are still further proved by the observations of Dr. Moore, which proved that measles is a disease of warm weather, or in other words, of that kind of weather which promotes the growth of the lowest forms of fungi and mouldiness.

Measles have been also shown to arise in connexion with musty linseed meal, by my friend and former colleague at Cork-street Hospital, Dr. Henry Kennedy, in a paper in the Dublin Medical Journal.

The only miasm which has as yet been shown to have any special connexion with scarlatina is that arising from the decomposition of slaughter-house refuse. This origin for scarlatina was first suggested by Dr. Carpenter of Croydon. Scarlatina has also been attributed to overcrowding ; but I have not yet been able to convince myself that the prevalence of scarlatina in connexion with overcrowding is to be attributed to any other effect of overcrowding than the well known tendency of that condition to favour the spread of contagion. To consider the question of the influence of slaughter-house refuse—Dr. Carpenter has shown in 9 cases of localised epidemics of scarlatina where the possibility of contagion seemed to be excluded, that the presence of decomposing slaughter-house refuse was the only assignable cause. It is unnecessary to give the details of these cases, but the most of them occurred under circumstances where all other sanitary arrangements were good.

The origin of scarlatina in connexion with decomposing

slaughter-house refuse, is further confirmed by an analysis of the death registry of No. 2 district of the south city district, undertaken by Dr. Maunsell, who found that out of 6,000 deaths registered in that district during the nine years 1864 to '72 inclusive, there were 268 deaths from scarlatina, of these 95, or more than one-third, occurred in the immediate neighbourhood of the slaughter-houses connected with the Clarendon, Castle, and Blackhall Markets, and another limited neighbourhood containing but one slaughter-house. The area in which these deaths took place is but one-eighth of the whole district. Two of these slaughtering districts are not remarkable for the prevalence of Zymotic disease.

The conditions which are essential to the production of whooping-cough are at present unknown, but the constancy with which it follows measles, points to the fact that what will control the latter will also control the former. Dr. Moore has shown how whooping-cough prevails in winter, measles in summer—the former following the latter and being aggravated by the effects of low temperature, favouring chest affections generally. No condition is yet known essential to the production of small-pox, but this is the most preventable of all Zymotic diseases by the simple and certain method of vaccination, so certain and safe a measure, that everyone is convinced of its certainty and safety except a few misguided and wrong-headed people who are more to be pitied than feared, and who should for the safety of society be handed over to the Commissioners who take care of the welfare of persons of weak intellect.

If other instances were required to show the value of sanitary measures, they could easily be produced. I will only mention two others as being derived from our own city. The cases of *trismus nascentium*, or nine days' fits, arising in infants in the Lying-in Hospital which have been I might say, annihilated by the preventive measures of ventilation and cleanliness first instituted by Dr. Clarke, and thus described in Dr. Churchill's able work on diseases of children :—

"No institution as far as I know has ever afforded such ample experience of the disease as the Dublin Lying-in Hospital, before the improvements in ventilation and cleanliness introduced by the late Dr. Joseph Clarke, to whom we are indebted for the best description of the attack. Dr. Joseph Clarke enumerates three especial existing courses of the disease—first, impure air ; second, neglect of keeping the infants clean and dry ; and third, irregularity of living on the part of the mothers, especially the abuse of spirituous liquors. At the end of the year 1782, of 17,650 infants born in the Rotunda Hospital 2,944 died within the first fortnight, or nearly every sixth child, and that owing to trismus. After the precaution he (Dr. Clarke) adopted the same pure and adequate ventilation in the hospital, out of 8,033 born alive, only 419 died in the hospital, or only 1 in 19½. During Dr. Collins' Mastership, of 16,654 infants born there were only 37 cases of trismus. Here is a splendid instance of the results of preventive medicine."

The other is the case of puerperal fever, a disease originating in the overcrowding of parturient women, as referred to in Dr. Farre's letter, quoted in the commencement of this lecture, which I think is very well shown in Dr. Evory Kennedy's work on this subject and the truth of which I believe has been fatally demonstrated in the Dublin Lying-in Hospital, but which, thanks to the reforms introduced by the late Dr. Collins, to the energetic efforts of the present Master, Dr. Johnston, and to the knowledge of defects pointed out by Dr. Evory Kennedy, is not likely again to afford the opportunity for demonstrating the dependence of this disease on bad sanitary arrangements.

Puerperal fever can, I believe, be almost if not altogether annihilated, like the nine days' fits, by isolation of the mothers either by separate buildings, as suggested by Dr. Kennedy, or by the complete isolation of the various wards by some other means.

Having shown the chief, original, and promoting causes of zymotic disease, it is manifest that the remedies are—

1. In building new towns or villages to select healthy sites.

2. Proper drainage, both house drainage and general drainage.

3. To prevent old ruinous and dirty houses from being inhabited, and to prevent new houses from being constructed so as to be injurious to the health of their inhabitants. Mr. Henderson will point out in his lecture how this is to be effected.

4. To prevent overcrowding either in houses or districts. This must be accomplished by constant inspection of all houses inhabited by the poor, by the regulation of the width of streets, the promotion of open spaces within towns, and by the breaking up of closed courts, and the making of wide thoroughfares through closed up neighbourhoods.

6. To promote cleanliness—1st, By the employment of all legal powers to compel and assist in the removal of dirt; and 2nd, To educate the people to believe that "cleanliness is next to godliness."

7. To provide proper accommodation for the sick at all times, and also during epidemics :—(*a.*) By proper hospital

accommodation at all times. (*b.*) By proper means of bringing patients to hospital. (*c.*) By the provision of special hospitals or wards, in connexion with general hospitals, to be used only in time of epidemics. (*d.*) Refuges where the healthy can be separated from the sick until the sick can be removed to hospital, and the houses or rooms they occupied cleansed and disinfected. (*e.*) The provision of accommodation for convalescents from zymotic diseases in convalescent homes. (*f.*) Proper and systematic disinfection of all places where sickness prevails or has prevailed.

These must be all accomplished by means of a well organised sanitary system, and I am sorry to say such a system exists in but few large towns, not at all, I may say, in the country, and scarcely anything worthy of the name of organization is at present to be found in Dublin. The treatment of the sick and the prevention of disease should be under the same department, which should also have under its control all matters for the relief of the poor, registration of births and deaths, and the performance of vaccination. Each large district should be under a Chief Medical Health Officer and every dispensary district should have for its Sanitary Officer the dispensary Medical Officer acting under the Chief Officer of the district. The Chief Officer should have almost absolute power, and should be only appointed with the consent of, and also removed by the chief sanitary authority of the state, namely, the Local Government Board. The absurdity of placing the administration of sanitary matters under the *absolute* control of Committees of Town Councils and Poor Law Guardians, many of them frequent offenders against sanitary law, is so great, that it will be at once perceived by every intelligent and thoughtful person.

I wish now to return my thanks to those who have given me their assistance in collecting materials for this lecture, namely, to Mr. Simon, the Medical Officer of the Privy Council; Dr. Burke, Medical Superintendent of the Irish General Registration Office; Dr. Ballard, of the English Local Government Board; Dr. Mapother, Medical Officer of Health for this city, and Dr. Maunsell, the able and energetic Secretary of the Poor Law Medical Officers Association. I have endeavoured to fulfil my difficult task to the best of my ability, and I trust that any shortcomings may be excused, and that you will believe that I have done my best to make a grave medical subject as little unpleasant and as interesting and useful as possible.

DIAGRAM IV.

Showing the Comparative Mortality from the seven principal Zymotic diseases during 9 years, from 1864 to 1872 inclusive.

in the City of Dublin.

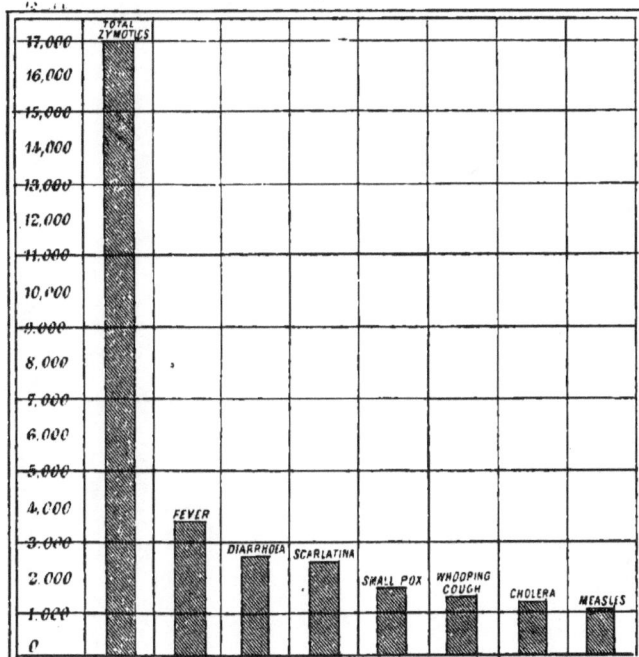

LECTURE VI.

ON LIABILITY TO DISEASE.

DELIVERED BY

ALFRED HUDSON, M.D., M.R.I.A.,

President of the King and Queen's College of Physicians in Ireland.

In his lectures on Public Health Dr. Guy observes that the science of Hygiene "makes application of a knowledge remarkable for its amount, and the great variety of sources whence it is derived." The truth of this observation has been illustrated by the gentlemen who have preceded me with reference to the sciences of chemistry, meteorology, and geography, and to the social condition and vital statistics of the community, and the relation of each to the genesis and diffusion of epidemics of zymotic disease. It now devolves upon me to occupy your attention for a short time with the consideration of those internal conditions and external agencies which increase our liability to disease, or, to use the language of the profession, act as predisponents or predisposing causes: and inasmuch as these conditions are either inherent in our constitutions, or involved in those surrounding agencies which minister to the nutrition of our bodies, or that molecular change, progressive and retrogressive, essential to healthy organic life, it will not be possible to explain their mode of action without reference to the laws which regulate this important function.

In the spread of epidemics two factors have long been recognised and illustrated by different comparisons. One of the best, says Dr. Hecker (in his History of the Epidemics of the Middle Ages), is the German word signifying "setting on fire" which compares the exciting disease in the *appropriate body* with the inflammation of combustible matter by the application of fire, or with the kindling of gunpowder by a spark. Another analogy employed by several writers compares the *materies morbi* to seed; the human body to the soil into which it is received, and concurring agencies to the seasonal influences which favour its germination, growth, and ripening, or reproduction.

This analogy corresponds to the three agencies which we severally denominate exciting, predisposing, and determining causes; by the first of which we denote the zymotic poison, by the second all those influences, intrinsic or extrinsic, which augment liability to disease; and by the determining cause anything which suddenly diminishes vital resistance to disease and neutralizes that conservative power by which noxious agents are assimilated and cast off from the blood; and thus determines the time and circumstances, but not the nature, of the invasion.

It has been held that "whenever predisposition and the specific poison are concurrently present the disease is invariably produced." We cannot admit the absolute truth of this dictum unless we deny the existence of that power which physiologists ascribe to the blood of assimilating and rendering innocuous, and eliminating noxious agents received into it from without; and we have moreover opportunities of recognising the influence of causes which while they suddenly depress the vital powers at the same time render active the germs of disease previously latent in the body. An example of no rare occurrence will illustrate this. An individual has been exposed to the contagion of typhus from day to day, but has hitherto assimilated and eliminated the poison received into his blood. He is brought into the immediate presence of a small-pox patient, and experiences a powerful feeling of fear or disgust, and forthwith sickens not with small-pox but with typhus. Here we have the seed and soil, in other words, the specific poison and liability, co-existing without disease being set up until a third influence intervenes, which we therefore denominate the determining cause. By thus recognising three factors we are enabled to explain not only such cases as that just mentioned, but also the immunity to contagion apparently possessed by individuals under long continued exposure, and those examples of sudden invasion under some special exposure recorded by Sir H. Marsh, Dr. Law, and other writers on fever.

It is no part of our present purpose to inquire into the sources or mode of action of contagion or to argue that it is a true ferment which being received into the blood excites in substances similar to those in which it originated changes identical with those which produced it, like thus following like, production and reproduction going on in a continuous series; or that it is a living germ having the power of multiplying itself in the blood of the infected person, or as has been recently argued "that a contagium particle is a detached portion of a diseased living body which coming in contact with a previously healthy body modifies the entire organism."*

It is not necessary for our present purpose to accept any of these explanations, and therefore in using the terms applied to any theory of contagion I wish not to be understood as adopting that particular theory.

There are however two facts in reference to the action of contagious zymotic poisons, it may be well to notice as bearing on our subject.

The first is that the liability to the special poison is exhausted in a greater or less degree by an attack, the liability to others being unaffected. This is particularly observable in regard to typhus and other eruptive fevers (exanthemata).

This fact, which has been somewhat differently explained by different physiologists, accounts for the much greater prevalence of epidemic diseases when introduced into virgin populations or those among which no previous epidemic of the kind had ever existed.

We have examples of such general liability in the histories of epidemics in populations unprotected by previous attacks, and under influences favourable to the spread of contagion. Such was the epidemic of measles which ravaged the Faroe Islands in the year 1846. It appears that the disease had not visited the islands for more than half a century; and that the ordinary rate of mortality of the islanders is very low; but it is stated by Dr. Parnum who investigated the epedimic in question, that it attacked scarcely less than 6,000 out of a population numbering between 7,000 and 8,000; few escaping except such as had suffered from the former epidemic or those who maintained a very rigorous isolation.†

Similar examples are found in the epidemics of small-pox in unprotected communities, more especially when introduced into America after the discovery of that country; three millions and a half of human beings having perished in Mexico alone. It has been stated that on its introduction into Canada 22,000 of the Red Indians were carried off by this disease, and that in Iceland in the year 1707 it destroyed 18,000 or more than one-third of the entire population.

Similar examples are recorded in the histories of the Black Death and sweating sickness of the middle ages. Of the former there died in London alone at least 100,000,

* Ross—" Graft Theory of Disease," chap. 2.
† *British and Foreign Medico Chirurgical Review*, vol. 7.

according to Hecker, and this writer estimates the mortality in Europe at 25,000,000.

The accounts given by historians of the habits of living and surroundings of most of the populations thus scourged, are such as fully explain their exceptional liability, and find their counterpart in the condition of the Mohammedan pilgrims during the last epidemic of cholera you lately heard so graphically sketched by Dr. Little, as also in that of the Irish peasantry in the famine of 1847-1848.

The other fact to which I have alluded is that each epidemic of contagious disease presents an uniform type, conformable to itself, not only in different places at the same time but also at different periods however remote, and that whenever exceptional forms and complications do occur these are due either to the co-existence of another epidemic or to special and exceptional forms of liability in the individual affected; the law being that with regard to the disease the poison has a special affinity or attraction for the organs or tissues according to its nature and the source from which it proceeded or was eliminated, as the skin and throat in scarlatina, the intestinal follicles and mucous membrane in cholera and enteric fever; and that with regard to the individual attacked those organs are found to become the seats of special complication which are specially predisposed in consequence of their undergoing increased disintegration or waste of tissue at the time of invasion of the disease.

We thus explain the increased liability to suffer cerebral and nervous complications in fever of the hard worked and anxious student or professional man, as well as many other complications not belonging to the disease *per se*, but to be ascribed to some condition in the individual affected.

Of the causes of liability or predisposition some are either inherent in our constitution, or in cosmical conditions not under our control, others being conditions in us or around us which are preventable and are therefore more peculiarly the subject of hygienic measures.

We may glance for a moment at some of the former before entering on the consideration of the others.

The first are those congenital or inherited constitutional peculiarities sometimes observed in an individual, sometimes in several members of a family, by virtue of which some persons appear to possess an absolute immunity from zymotic disease under any amount of exposure, while on the other hand, others succumb to its influence whenever exposed.

At a recent meeting of one of our Medical Societies, the case was mentioned of a lady whose liability to smallpox was such that she had suffered seven attacks of this disease. Nor is this a singular instance.

Different members of a family not unfrequently share in this kind of susceptibility, and the records of medicine moreover contain numerous examples of a family predisposition to suffer some unusual or exceptional complication in the course of typhus or other zymotic disease. Of course these inherent and inherited forms of liability cannot be prevented, neither can the family tendency to certain complications be explained, farther than by supposing that there exists in such individuals not only conformity of type of structure, but also a conformity in those "affinities existing between definite tissues and definite substances, which must be referred to peculiarities of chemical constitution in virtue of which certain parts are enabled in a greater degree than others to attract certain substances from the neighbouring blood." *

Another non-preventable form of liability is that attendant upon the evolution of organs, and the rapid metamorphoses of tissue during the growth of the body. That the liability to certain diseases both diathetic and zymotic is most remarkable during the periods of childhood and adolescence is well known. We find the explanation of the fact, so far as it can be explained, in the greater activity of the formative process and consequently great necessity that the balance of forward and retrogressive change should be preserved, and moreover that the balance of evolution of the several organs should be so adjusted that none of the materials appropriate for the maintenance of any part may remain in the blood; "since each part, by taking from the blood the materials it requires for its nutrition, prevents the accumulation and excess of such matters in the blood as effectually as if they were separated from it and cast out by the excreting organs specially provided for that purpose." †

It follows of necessity that the period of growth and development is one of general liability, capable of being augmented by various agencies, extrinsic or intrinsic, till it amounts to predisposition or proclivity to disease.

Another and remarkable example of liability, is that caused by the involution and disintegration of an organ which has fulfilled a temporary purpose in the economy, and the consequent presence of its effete materials in the blood. Such is

* Vinhow, Cellular Pathology, page 123.
+ Kirke's Physiology, page 95.

I

the case of the puerperal female whose liability to zymotic disease is well known.

Dr. Moore has already illustrated the influence of season and of atmospheric changes as *exciting* causes of disease; they have also an influence on our organism, predisposing to different types of disease as well as to different diseases at different periods.

The researches of Dr. Edward Smith and other physiologists have shown that great variations in the vital processes occur at different periods of the year, and that an exaggeration of these constitutes a form of liability to disease, and impresses a type upon disease, varying according to the season, "the tendency generally being to sthenic and inflammatory forms of disease in spring when the amount of vital action is at a maximum, and to those of an asthenic type characterized by exhaustion in the latter end of summer and in early autumn." "It appears, however, that the rule, though generally true, is not universally so, but that the effects of season are modified by the constitutional peculiarities of individuals, and that the selection of certain individuals as the earliest victims of an epidemic of influenza or cholera is due to the comparatively greater influence which certain external conditions exercise upon the vital powers of the system in these individuals."*

There cannot be a doubt that the cycle of change of the seasons is adapted by the all-wise Creator to our bodily and mental constitution, and that any marked deviation from their order must be injurious, whether it be the unusual prolongation of a season, or its abbreviation, or the excess or defect of its characteristic phenomena. Thus, prolonged wet or drought, prolonged cold and frost, or prolonged summer heat, are each injurious. Warm and moist winters are proverbially unhealthy, as are cold, dark, and ungenial days, with deficiency of sunlight, in spring. The last especially predispose to typhus, while long, hot summers not only favour the generation of the malarial poisons, but moreover render the body predisposed to their reception and influence, more especially when passing suddenly into a cold and moist, because late, autumn.

The last non-preventable predisponent I shall notice is that mysterious atmospheric agency which medical men recognise by the terms epidemic constitution, epidemic influence, or, in the words of Inspector-General Lawson,†

---

* Health and disease as influenced by the cyclical changes in the human system, chap. 6.

† Army Medical Reports, 1861, page 405.

designate " Pandemic wave," " a series of waves generated
in southern latitudes which flow to the north or north-
westward in succession, leading to an increase of fever at
every point over which they pass." "It must be admitted,"
says Dr. Lawson, "that as in different countries different
forms of fever prevail under the same general influence the
pandemic cause determines the frequency and severity,
rather than the particular form of the fever, which there are
many reasons to conclude is more intimately connected with
the local circumstances at the time. . . . It is characteristic of a pandemic wave that during its passage local
causes which, under ordinary circumstances, seem to exercise
inconsiderable influence over the health of those exposed to
them, then display a potency which, if regarded without
due weight being given to the reigning pandemic influence,
seems quite unaccountable."

Of the existence of some such general cause or influence,
and of its power to stamp its peculiar features upon prevailing zymotic disease, or to cause certain exceptional complications to occur during the period, no careful observer can
entertain a doubt; but neither can we offer an explanation
of the *modus operandi* of this occult and mysterious influence or suggest any measures by which it can be neu-
tralized or prevented.

Of those predisposing causes, more important in a hygienic
point of view, which are preventable by suitable precautions on the part of individuals or the community, it has
been remarked by Dr. Carpenter "that they all tend to
introduce an accumulation of disintegrating azotized compounds in a state of change in the circulating current," and
are all reducible to three categories:—*

I. Those which tend to introduce into the system decomposing matter that has been generated in some external
source.

II. Those which occasion an increased production of decomposing matter in the system itself; and,

III. Those which obstruct the elimination of the decomposing matter normally or excessively generated within the
system, or abnormally introduced into it from without.

It has been observed that liability to disease in a general
sense is a law of our being, involved so to speak in the
function of nutrition, or that function by which is effected
the continual progressive and retrogressive change of the

* *British and Foreign Medico Chirurgical Review,* vol. xi.

particles of our bodies; each organ or tissue attracting from the blood the materials adapted to its own growth or maintenance, while the products of retrograde metamorphosis— in other words the effete materials of the tissues—are resorbed into the blood, and by combination with its oxygen form new and devitalized compounds fit only to be eliminated or cast off by the various excreting organs; the carbonized products being chiefly eliminated by the lungs and liver, the azotized by the kidneys and other emunctories.

It follows that this process of continual molecular change necessarily involves the temporary presence in the blood of a variable amount of matters in a state of progress to decay, the elimination of which is essential to health.

We know that this elimination and the preservation of health depends mainly on the healthy condition of the three factors of nutrition, the blood, the tissues, and the nervous system;[*] the first requiring healthy digestion, healthy respiration, and healthy secretion, for the preservation of that assimilating power it possesses, by which many noxious substances introduced into it from without are changed and made harmless, and ultimately eliminated; the second being equally necessary, inasmuch as the unhealthy tissue reacts upon the blood, furnishing oftentimes a permanent supply of noxious ingredients upon which a *dyscrasia* or blood disease depends.[†] The necessity of a due supply of nervous influence for the nutrition of the body or of any portion is proved by numerous examples,[‡] as is also the important part in predisposition played by the exhaustion of this influence by excessive and long continued exercise, whether mental or bodily, anxiety, or other depressing emotion on the one hand, and the protecting power of the opposite condition of mental and physical energy and activity upon the other.

In short, it will be found that so long as the functions of these several factors of nutrition are duly performed, and coadapted to each other, so long the equilibrium of health is preserved, and the reaction against morbific agencies is maintained; but if the health of the blood suffers by contamination from within the body or from without, to the loss of its assimilating power; or the healthy metamorphosis of the tissues is interfered with; or an important excreting organ fails to exercise its depurative functions; or by some severe shock or prolonged strain the nervous influence is per-

---

* *Vide* Sir James Paget's Lectures on Surgical Pathology, lectures 1 and 2.
† *Vide* Virchow's Cellular Pathology, lecture 6.
‡ *Vide* Paget, lecture 2.

verted or lost;" the continuous adjustment of internal relations
to external relations " in which healthy life consists* becomes
imperfect, the reaction against external morbific agencies is
no longer maintained; to the receptivity of these which
exists in all constitutions is added the incapacity of assimilating and changing so as to eliminate them by excretion, in
which proclivity or predisposition to disease essentially consists, and the zymotic poison being introduced from without
—whether in the form of contagium particles detached from
living diseased bodies, or of miasm emanating from other
sources—its special dynamic effects are set up in the blood,
and there follows disease conformable to its source or type,
with local affections due either to the elective affinities of
the poison for the organs or tissues in which the contagium
is generated, or to pre-existing conditions of the organs thus
affected; and, finally, the reproduction and diffusion of the
contagium.

This view of the essential nature of predisposition corresponds with the explanation given by Dr. Carpenter of the
mode of action of predisposing causes, viz. :—

"That all the recognised predisposing causes of zymotic disease
tend to produce in the blood an undue accumulation of azotized
matter already in a state of retrograde metamorphosis, and therefore
precisely in the condition in which it is most readily acted on by
ferments, . . . and that the liability of each individual among
a number who may be concurrently exposed to the same poison will
mainly depend upon the degree in which his blood may be charged
with the matters in question."

Dr. Carpenter's theory not only explains the action of
causes predisposing to disease generally, and to special complications in particular, but it also explains the injurious
influence of the latter upon the disease, and the well known
fact that the fatal termination of many cases of fever and
scarlatina is due to continuous infection arising from the
reception into the blood of the products of regressive change ;
constituting what are denominated by pathologists the
secondary contaminations of febrile diseases.

Under his first category of predisposing causes Dr.
Carpenter classes putrescent food, water contaminated by
decomposing matter, and contaminated respired air.

Of the first no more striking example could be cited than
that of the Faroe Islanders already referred to, whose extra-

* II. Spencer's Biology, vol i.

ordinary liability to zymotic disease is to be accounted for by their diet, thus described by Dr. Parnum :—

" During the many months that the fish, flesh, or fowl is neither fresh nor yet wind dried it is called 'rast,' a word which I can only translate by half-rotten. This appellation it fully deserves from the horrible smell that it sends forth, from its mouldy aspect, and the numerous maggots which swarm upon it. I have seen a boat's crew of eight men eating with relish the raw flesh of the ca'ing whale, though it was so decomposed that the smell of it was disagreeable to me even in an open boat, and the bottom of the boat was almost white with the maggots that fell from the decaying mass."

We can scarcely wonder that on the introduction of the zymotic poison of measles nearly six-sevenths of the population were attacked, or that a most exhausting diarrhœa, often continuing for months, was a frequent sequel of the disease.*

Of the influence of contaminated water as a predisponent to cholera and enteric fever, and as powerfully augmenting the severity of these diseases, the proofs are so numerous and so well known that I need not dwell on them ; besides you have already heard from Dr. Grimshaw abundant evidence of the intimate causal relation between this agent and zymotic diseases.

The same remark applies to the respiration of the gases arising from decomposing human excreta or " civic miasm," but just as there are some individuals who, because contagion does not affect *all* who are exposed to its influence, do not be-

---

* It should be remarked that the evil effects of the excessive use of animal food are by no means exclusively confined to that in a state of putrescence. It is, however, not so much as a predisponent to zymotic disease as the cause of serious and often fatal complications that the influence of a too highly animalized diet is observed.

The dangers of a state of rude health produced by full living have been thus graphically sketched by the late Mr. Travers:—

" The state of rude health, as that phrase is commonly understood, I consider to be a forced state, that in which the nutrical powers are tasked to the uttermost and successfully struggle with a surplus of diet and stimulus, ridding the body of both by the action at its full stretch of every excreting organ. The subjects of this class are perpetually running upon the boundary between health and disease, a sudden shock deranging some important function destroys the equilibrium of the machine which its over pressed powers are the less capable of reinstating," &c.

Except in regard of nomenclature, there seems little difference between the teachings of a great surgeon of the last half century and those of the physiologists of the present day.

I find predisposition to epidemic disease ascribed to the excessive use of animal food by Sims, in his description of a remarkable epidemic of typhus in 1771, and by Hecker in his account of the sweating sickness of 1517, which prevailed exclusively among men, " who," says Hecker, " eat spiced meat to excess, but who were also addicted to nocturnal carousings, and drank strong wine on rising in the morning."

lieve in its existence—at least as an essential element in the causation of disease—so there are others who, because individuals or families live for months and even years in the midst of filth, and habitually breathing air loaded with fœcal miasma with apparent impunity, question the influence of such miasm as a source of the poison of fever, or exciting cause of the disease.

The answer to this objection is, that immunity under such circumstances is owing to the power which the blood possesses of assimilating and rendering innocuous noxious matters received into it from without. It is this power of adaptation or acclimatization which enables the inhabitant of the Faroe Islands to live upon putrefying flesh without suffering anything more than ordinary diarrhœa unless during epidemic visitations; which accounts for the fact that the native of tropical malarious districts can reside in the vicinity of the swamp which is fatal to the European stranger; which long made French physicians regard the enteric fever of Paris as a disease peculiar to the strangers visiting that city; and which enables many medical men, nurses, and hospital servants to pass years in the midst of infection with impunity.

But whatever opinion may be entertained with regard to civic miasm as an exciting cause, its influence as a predisponent cannot be questioned, both as rendering the constitution more liable to zymotic diseases in general, and as determining their special complications. The histories of the epidemics of zymotic diseases of the present day abundantly prove this, as do the descriptions given by Hecker and others of the abounding filth of the persons and dwellings of the English during the epidemic periods of the middle ages, and the evident causal relation between the diseases and these conditions.

In such cases we frequently find that a period of deranged health precedes the outbreak of fever, characterized by gastric derangement, headache, languor, and unhealthy secretions; and observations made by medical men at different times and in different places have shown that in this premonitory stage the blood undergoes a change recognisable by the microscope, the red corpuscles being broken up and the serum tinged by their contents.*

Dr. Grimshaw has adduced abundant proof of the power

---

* This fact, first observed by Dr. Potter of Baltimore, and again by Dr. Cormack, in the Edinburgh relapsing fever of 1843, has been more recently recognised and described by Dr. Hand of Philadelphia. Dr. Hand relates a case in which the blood was examined within three or four hours of the first seizure and found to be in an average state of degeneration. The case proved a typical one of relapsing fever. Dr. Hand believes that by living in a contaminated atmosphere the blood

of *ochlesis,* or the poison generated in crowded collections of human beings, with insufficient supply of pure air, to generate and diffuse the contagium of typhus. That this poison may be generated *de novo* under certain conditions I entertain no doubt, but that ochlesis more frequently acts as a predisponent to zymotic diseases in general, and to exanthematous typhus more especially is probable.

Examples are to be found in the histories of the famine fever of 1847 and 1848 detailed in Sir William Wilde's able and elaborate Census Report of that period, they are also to be found in the cholera reports of the Board of Health and Privy Council, and in the Army Medical Reports of cholera in India.

Perhaps no more striking example of its *preventable* nature could be cited than that afforded by the *trismus nascentium,* formerly so fatal to newborn infants in the London workhouses, and in the great lying-in hospital of this city, and still so in Iceland.

In the lying-in hospital the deaths within the first fort-night after birth formerly amounted to 1 in every 6 children. The improved system of ventilation adopted by Dr. Joseph Clarke reduced this proportion to 1 in 19½. Under Dr. Collins' mastership this was reduced to 1 in 450, and under the present master, Dr. George Johnston, it has been further reduced to little over 1 in 500.

In Iceland, under opposite hygienic conditions, we learn that during twenty years ending in 1847, 64 per cent. of the infants born alive died of *trismus* from the fifth to the twelfth day after birth.

The immediate effects of overcrowding and deficient ventilation are (*a*) diminished proportion of oxygen in the respired air, (*b*) diminished oxidation of tissue, and consequently continually increasing amount of decomposing nitrogenous matters in the respired air and in the blood; but in times of war and famine other conditions co-operate. With regard to the latter, Dr. Carpenter observes :—

" We have not merely that general depression of the vital powers which is a predisposing cause of almost any kind of malady, and pre-eminently so of zymotic diseases, but also the presence of a large amount of disintegrating matter in the blood and general system which forms the most favourable nidus possible for the reception and multiplication of such poisons. And thus it hap-

may become thus changed without fever necessarily following. This was noted in the case of four of the resident medical attendants of the Philadelphia Hospital.— *New York Medical Journal*—quoted in *British and Foreign Med. Chirurgl. Review,* Oct., 1870.

pens that pestilential diseases most certainly follow in the wake of a famine, and carry off a far greater number than perish from actual starvation."*

Such was the case in the Irish famine of 1846-7 the effects of which are thus graphically described by an eyewitness—Dr. Donovan of Skibbereen. "'In a short time the face and limbs became frightfully emaciated; the eyes acquired a most peculiar stare; the skin exhaled a peculiar and offensive fœtor, and was covered with a brownish filthy coating, almost as indelible as varnish. This I was at first inclined to regard as incrusted filth, but further experience has convinced me that it is a secretion poured out from the exhalents on the surface of the body. The sufferer tottered in walking like a drunken man; his voice became weak like that of a person in cholera; he whined like a child, and burst into tears on the slightest occasion. As regards the mental faculties their prostration kept pace with the general wreck; in many a state of imbecility, in some almost complete idiotism,'" &c.†

The highest degree of predisposition of which the living body is susceptible is generated by the crowding together in gaols, workhouses, or hospitals of human beings in the condition above described, as is proved by the annals of the epidemic of 1847-8 already referred to.

Fatigue and exhaustion caused by prolonged or excessive bodily and mental exercise is one of the most powerful predisponents to zymotic disease. We see its influence in the persons of anxious relatives who succumb to infection when exhausted by long watching; and it is witnessed in the case of cholera in soldiers exhausted by long marches in a hot climate, more especially if conjoined with intemperance in the use of spirits. Under such combined conditions we have increased disintegration of muscular tissue, involving an increased accumulation in the blood of carbonized and nitrogenous products in a state of progress to decomposition, and diminished oxidation of effete tissue due to imperfect respiration, and to the superior attraction of the alcohol for oxygen—heat and alcohol here acting under Dr. Carpenter's third catagory, namely, by obstructing the elimination of the products of disintegration of muscular tissue already augmented by excessive exertion.

We are indebted to the Rev. Professor Haughton for the remarkable observation that the amount of nitrogenous products of metamorphosis of tissue from mental work exceeds that accruing from bodily labour in the proportion or 533 grains to 400 grains of urea excreted daily.

* "Principles of Physiology," sixth edition.
† *Dublin Medical Press,* vol. xix, page 67.

He also states that "whenever an abnormal amount of
these products is excreted the cause must be ill health; and
most generally that most fatal of all diseases to which man
is liable, anxiety of mind, a vague and unscientific expression," says Dr. Haughton, "which, however, denotes a real
disease."*

There may be said to be two distinct modes of action of
mental, or more properly speaking emotional, predisposing
causes, viz., by long continued strain and by sudden shock.
Carking care, anxiety, and despondency act in the former
manner as true predisponents; sudden and violent emotions
of grief, shame, and terror act in the latter, and may be
more properly termed determining causes.

Both exercise a powerful influence on the molecular nutrition as we have already seen in the case of the first; a
striking example of the influence of terror on this function
is thus narrated by Mr. Carter:—"A lady who was watching
her little child at play saw a heavy sash fall upon its hand
cutting off three of the fingers, and she was so much over-
come by fright and distress as to be unable to render it any
assistance.   A surgeon was speedily obtained, who having
dressed the wounds turned himself to the mother, whom he
found seated moaning and complaining of pain in her hand.
On examination, three fingers corresponding to those injured
in the child were found to be swollen and inflamed, though
they had ailed nothing prior to the accident.   In four-and-
twenty hours incisions were made into them, and pus was
evacuated, sloughs were afterwards discharged and the
wounds subsequently healed."†

It is well known that under the influence of strong
emotion the blood and secretions will undergo important
changes, the surface will become pallid, or it may be suffused
with bile, the mother's milk will become a deadly poison to
the infant at the breast, and in many cases of fever recorded
by Sir H. Marsh, Dr. Law, and others, in which such emotion
has been the *determining* cause the entire course of the
disease has been characterized not only by severe nervous
symptoms but also by marked changes in the blood, and
lesions of the function of nutrition.

The conclusions which I think we may draw from the
facts which I have adduced, are the following:—

I. That liability to zymotic disease is inherent in our
constitution, involved so to speak in the function of nutrition.

* *Dublin Quarterly Journal of Medical Science*, vol. 30.
† On Hysteria, quoted by Dr. Carpenter.

II. That it varies in degree in different individuals, and in the same individual at different times and under different conditions, partly external or extrinsic, partly internal or intrinsic, some of which are preventable and others non-preventable in their nature.

III. That, *ceteris paribus*, this liability is least in those persons in whom healthy blood, healthy tissues, and healthy excretions, and a healthy state of the nervous system constitute a healthy nutrition.

IV. That is greatest in those whose blood contains the largest amount of the products of waste of the tissues, or of matters in a state of decomposition introduced into the circulation from without.

V. That all scientific hygienic measures are based upon their power of preserving or restoring the healthy condition of the factors of nutrition and neutralizing the conditions, whether extrinsic or intrinsic, by which this function is impaired or deranged.

LECTURE VII.

ON ANTISEPTICS AND DISINFECTION.

DELIVERED BY

ROBERT MACDONNELL, Esq., m.d., f.r.s.,

Surgeon to Dr. Steevens' Hospital.

THE impurities existing in the atmosphere which surrounds us are partly gaseous, and in part minute but solid particles of matter.

The gaseous impurities which render air more or less deleterious, such as the carbonic acid gas which accumulates in a crowded room, or the sulphuretted hydrogen which emanates from the sewer are detected by chemical agents. Chemistry has taught us how to recognise these impurities and how to remove them. Many gaseous impurities in the air we detect by the sense of smell, but some being inodorous can only be proved to exist in it by chemical re-agents. When, therefore, we remove disagreeable smells, and so far purify air as to cause it no longer to be offensive to our nostrils, it by no means follows that it is thereby rendered healthful and pure.

It is a dangerous delusion to repose trust in that class of agents called " deodorants"; they are very useful in their way when their true use is comprehended ; in so far as they render air less disagreeable to our noses they are good ; if thereby we are induced to suppose that the air is purified and rendered wholesome we are led into error. It must ever be borne in mind that no chemical disinfectants can supply the place of cleanliness, ventilation, and drainage.

But it is not to the gaseous impurities of the air we breathe that I desire at present to direct attention. It is to the other class of impurities of which I have spoken : the exquisitely minute but solid matter in suspension in the air ; the almost inconceivably fine dust which dances in the sunbeam, and is borne across the ocean by the storm.

The offensive gases given off from decomposing organic matter are usually either ammonia or compounds of hydrogen with carbon, sulphur, or phosphorus. Such impurities as these are absolutely invisible. Thanks to the marvellous delicacy of the means of investigation introduced through the influence of light, we are now enabled to render visible the dust, minute though it be, which floats in our atmosphere.

The dust, or if you will, the dirt of the air, is of course very complex and very varied in its composition; it is not exactly alike in the flax or cotton mill and over the threshing machine, in the crowded theatre and the hospital ward. It is however everywhere composed of the débris of organic as well as inorganic matter; but along with this pulverised débris of lifeless matter there is reason to believe that there exist minute spores, germs, or seeds, which being borne by the gentlest draft of air from place to place, are as capable of germinating and growing as the thistle down or the dandelion seed, if chance bears them to a suitable soil.

According to an analysis made by Dr. Percy, the dust collected from the British Museum contains fully 50 per cent. of inorganic matter. No doubt most of this is mineral matter, worn off from the streets and houses of London, and matter carried forth from factory chimneys. The débris thrown off in the cotton mill, or by the gradual wear and tear of our clothes, carpets, the surface of our bodies, or at the brushing of our hair, &c., &c., &c., is organic matter, but it is lifeless; it is not capable as the minute germs or spores are of growing, and so of calling into play any of that remarkable series of changes which we are familiar with, as accompanying the growth and development of such plants as the yeast plant or vinegar plant.

Although, then, these minute seeds, germs, or spores form, in all probability, a very small part of the atmospheric dust compared to the lifeless organic or inorganic débris, they are, nevertheless, on account of their active properties, and of the wonderful changes (fermentations, decompositions), they are capable of calling into existence, in certain fluids beyond all comparison, the most important and influential agents present in the dust of the air.

The lecturer now exhibited to the audience the well-known experiments of Professor Tyndall, showing how large a portion of the dust of the air is really of organic origin. In an electric beam, which powerfully illuminated the dust of the theatre, an ignited spirit lamp was placed. Above and around the flame were seen wreaths of darkness resembling an intensely black smoke. But this blackness was proved not to be smoke, for a similar blackness was produced by a hydrogen flame from which no carbon could pass away, and a red hot poker placed beneath the beam gave rise to a similar phenomenon. Moreover, when real smoke was allowed to rise across the beam, so far from giving rise to wreaths of darkness, it caused clouds of snowy whiteness.

The darkness then is not smoke, it is simply that of

stellar space: the organic particles floating in the beam being destroyed by the heat, there is no longer anything to catch and reflect the light. The vacant space is darkness, rendered visible by contrast.

Having exhibited a variety of experiments, as set forth by Professor Tyndall, to illustrate this subject, the lecturer proceeded to say :—After such evidence as is now before us no one can doubt the large quantity of organic filth which, in the shape of dust, loads the atmosphere of cities, nor is the country free from its pollution. Even far out at sea, and on the summits of mountains, these light bodies may be met with.

A phial of perfectly pure, newly-fallen snow, was taken from the summit of Mont Blanc, by Dr. Kolbe, and brought to M. Pouchet. On melting it it yielded about one cubic inch of water, which was to all appearance pure and clear. But a slight deposit was observed in it on standing, and this deposit contained the following substances :—A few minute bodies of a mineral nature, two woollen filaments, one white and one blue, a fragment of a confervoid plant, a minute tuft of vegetable air tubes, and a dozen young cells of Protococcus Nivalis. Thus we find that the force of the wind may bear, even to the Alpine summits, dust containing mineral matter, organic matter, and spores.

As regards the " germ theory " of disease, and as it may be called the " germ or putrefaction theory " of suppuration, science owes much to discussion arising upon a very different topic—viz., that of " spontaneous generation." Two able disputants arose in M. Pasteur and M. Pouchet, whose investigations, although undertaken with quite another object, have yielded a rich harvest in this field.

M. Pouchet declared that all his examinations showed the atmosphere to be everywhere poor in organic germs, and often entirely destitute of them; and that its capacity for generating animal life resided not in these germs, but in the general vivifying power of the air. M. Pasteur, on the other hand, insisted that the chemical constitution of the air remaining the same, its power of producing organic life varied with the locality from which it was taken ; and this because the number of germs contained in it varied in different places.

Both the disputants stated their positions in definite terms. M. Pouchet said, " I assert that from whatever region of the globe I take a quantity of atmospheric air, if this air be placed in contact with a putrescible liquid in hermetically-sealed vessels, the liquid will invariably become filled with living organisms."

M. Pasteur said, " It is always possible to obtain in a particular locality a notable volume of atmospheric air which, without having been subject to any physical or chemical modification, is nevertheless incapable of exciting any change whatever in a putrescible liquid." These assertions, emanating from two eminent observers, both members of the Academy, were so diametrically opposed to each other, that it was agreed to refer them to a Committee, in whose presence the requisite experiments should be performed, and who should report to the Academy on the result. Such a Committee, composed of five members, was accordingly formed, and entered upon its labours in June, 1864, in the Chemical Laboratory of the Museum of Natural History, at the Garden of Plants.

M. Pasteur first presented three of his flasks which had been filled with air four years previously on the Montanvert, and had remained ever since perfectly unchanged. One of them was opened under mercury, and the air which it contained, on being analyzed, was found to have the natural constitution of the atmosphere (twenty-one parts of oxygen to seventy-nine parts of nitrogen). Another flask was opened by a minute orifice at the neck, and after being left for three days exposed to the atmosphere, it contained flakes of a cryptogamic vegetable growth, which subsequently became largely developed.

M. Pasteur then prepared and sealed, before the Committee, sixty flasks, similar to those previously used. Nineteen of them, after cooling, were opened and immediately resealed in the amphitheatre of the Museum; nineteen on the top of the dome of the same building, and eighteen others at a country-house a few miles from Paris, under a thick growth of poplars. Afterwards microscopic vegetations were developed in five flasks of the first set, six of the second, and sixteen of the third. All the remainder were unchanged at the end of over four months.

The Committee subsequently reported the result of their experiments, and gave as a conclusion that the facts observed by M. Pasteur, and contested by M. Pouchet, were of the most absolute exactitude.

It thus seems to have been placed beyond a doubt that the atmosphere is incapable, from its chemical constitution alone, of exciting organic growth in a boiled infusion ; but that it often introduces with it into the solution invisible germs which have this effect, the proportion in which these germs are present varying with the locality from which the air is derived.

But up to this time the dispersion of organic germs in the

atmosphere was not an actually observed fact, but only a probable inference from the results of experiments like the above. This is what gave a certain weight to the objection of M. Pouchet when he said in one of his communications, " It seems to me that when an experimenter declares that he can collect from the atmosphere either the eggs or spores of microscopic organisms, we have a right to demand that he should show them to us."

No one, in fact, had succeeded in collecting these germs from the air in any abundance, in such a form as to be visible and recognised.

This, however, was accomplished by Dr. Lemaire in 1864. He adopted the plan of condensing the vapours of the atmosphere in glass tubes by means of artificial cold. The moisture thus obtained was then kept in the tubes, well stoppered, together with an equal or double volume of air, at a temperature of from 73° to 86° Fahr. The collections were made in the month of July, from a marshy neighbourhood in the country, from the Garden of Plants in Paris, and from a village near the city, situated at two or three hundred feet higher elevation. The liquid, when first condensed, was colorless and limpid. It contained microscopic vegetable germs or spores ; a great number of pale cells, of different dimensions ; a considerable abundance of very small semi-transparent bodies (thought to be the germs of future infusoria) of a spherical, ovoid, or cylindrical shape, sometimes regular and sometimes irregular ; certain brownish corpuscles, apparently of vegetable origin ; starch-grains, dust-particles, and cubical crystals. Within twenty-four hours afterwards there were developed an abundance of living infusoria, bacteria, vibrios, spirilla, and monads, together with ramified cryptogamic vegetations. Exactly in proportion as the cryptogamic vegetations and the infusoria were developed, the spores and the small semi-transparent corpuscles were found to disappear.

Thus the actual existence of organic germs in the atmosphere was demonstrated ; and there could no longer be any doubt that these germs, when introduced into an organic infusion, are abundantly sufficient to account for the production of infusorial and vegetative life.

The Lecturer, in order to enable his audience more accurately to comprehend the exact form of M. Pasteur's experiment, exhibited the mode of hermetically sealing up flasks containing fermentable liquid.

M. Pasteur took glass flasks filled partially with a clear infusion of brewer's yeast. He then boiled the fluid, and while ebullition was going on actively drew out the necks of the flasks to a narrow point and sealed them over the flame of a blow-pipe. The fermentable liquid was thus enclosed in an air-tight vessel containing nothing save its own rarified vapour. Upon cutting off the neck of the flask in any particular place, the air of that place rushed in to fill the vacant space. The flask being then resealed the effect of this air and (of such germs as it might happen to carry in along with it) upon the liquid could be observed.

He prepared sixty of these flasks. Twenty of them were afterwards opened and resealed in the country at the foot of the first plateau of the Jura range. Twenty others were opened and resealed on one of the Jura mountains, two thousand five hundred feet above the level of the sea, and the remaining twenty near the " Mer de Glace " Glacier, at an altitude of six thousand feet. The result was that of the first twenty flasks, eight were found afterwards to have produced living organisms ; of those filled with air from a point two thousand five hundred feet above the sea level, five showed similar productions ; while of those filled at the " Mer de Glace " one only became the seat of organic life.

We see then that the air of our large cities is loaded with dust in part either mineral or lifeless organic débris, but partly composed of matter in the shape of germs or spores capable under certain circumstances of starting into organic life and growth. Nor is the air of the country or the mountain top quite free. Although ordinary light permits this dust to escape our observation, a strong beam causes it to become a real visible existence ; painfully real when we come to contemplate with the aid of the electric beam the fine filth which we every moment draw into our lungs. It is, however, quite certain that air so laden with dust of one kind or another as to be positively irritating to the air passages, and even capable of gradually developing diseases peculiar to certain trades is not necessarily charged with that kind of matter " living dust," as Professor Lister aptly calls it, which is so very deadly to man ; which in fact is conceived in the present day to be the means of propagating epidemic disease. The question is what is this portion so virulent in its nature, so mysterious in its development, so terrible in its attacks upon mankind ? The current theory some time ago concerning epidemic diseases was that they were propagated by a kind of malaria consisting of organic matter in a state of motor-decay, and that such matter entering the body spread there the destroying process which had attacked itself. This theory was exactly analogous to that held with regard to the supposed action

K

of yeast. A little leaven leavened the whole lump. The discovery made by Cagniard de la Tour in 1836, and independently by Schwann of Berlin in 1837, altered the views of chemists with regard to the theory of fermentation, and gradually altered the view hitherto taken as to the causation of epidemic diseases. By the discovery of the " Yeast plant," a living organism capable of feeding, growing, reproducing itself, fermentation was proved to be a product of life, not a process of decay ; a decomposition if you will, but a decomposition caused by the energy of growth and life. As regards fermentation the minds of chemists, influenced by the authority of Gay-Lussac for a time ascribed putrefaction to the action of oxygen, and retained the idea of matter in a state of decay. Pasteur however finally exploded this notion. He proved that the so-called ferments are not such; but that the true ferments are organized beings which find in the reputed ferments their food. Side by side with these researches and discoveries concerning ferments and fermentation has run the "germ theory" of disease. It is true that it is in a great degree an hypothesis based upon analogy, and every philosopher will admit that in such matters analogy may be but a deceitful guide. It has however received much strengthening and support from various scientific observations. Unconsciously perhaps the mind is prepared to accept such a theory by learning the strange history of the invasion of the minute entozoon known as the "Trichina spiralis." Dr. J. Burdon Saunderson's experimental inquiries relating to the nature of infective agents has done much in the same direction. The researches of this able observer go to show that in all infective inflammations in the lower animals microzymes (microscopic organisms) abound in the exudation liquids ; and that the same forms are to be found in the blood of animals when in the state of acute infective fever.

To turn, however, from these refined scientific observations and ingenious theories to their practical application, the question is, how can we best escape the dangers which beset us on every side from the living organic impurities of the atmosphere ? Obviously by seeking to free the air from these impurities.

As to " *deodorants*," the risk of trusting to them simply has been already spoken of, they have, however, a recognised value. Various kinds of charcoal (peat charcoal, boghead coke, &c.), quicklime, chloride of lime, a variety of metallic salts, dry earth, &c., have the power of either removing by their absorbing power offensive gases or of breaking them up by their chemical action. Ammonia, and several of the compounds of hydrogen, with carbon, sulphur, or phosphorus, not unfrequently given off by decomposing animal matter, may thus be got rid of.

True disinfectants or oxydizers of organic matter are more valuable as well as safer. The fumes of nitric and nitrous acids, the manganates and permangates of soda and potash, chlorine gas, &c., are powerful oxydizers, and quick lime, and chloride of lime owe in fact their deodorizing qualities to the same cause.

Antiseptics, or those bodies which restrain or absolutely prevent decomposition, are in many respects the most important. Disinfectants oxydize the products of decomposition; antiseptics prevent the formation of any such products. The ordinary processes of cooking, pickling, tanning, &c., are to some extent antiseptic processes. Various metallic salts, sulphurous acid, creasote, and carbolic acid have remarkable antiseptic powers. Small quantities of the last named acid added to an organic solution completely prevent the growth of those organisms which cause, or at least accompany, decomposition. The careful use of it has been found to produce the best results in the treatment of open wounds. It seems to kill the "living dust," which, penetrating along with the air into open wounds, causes the blood to rot and the system to be infected by a poison thus generated within itself.

The whole range of modern scientific research does not possibly offer a more charming illustration than this topic of the antiseptic treatment of wounds of how science and practice work hand in hand for the benefit of mankind.

When the philosophers of Bologna discovered and investigated the elementary phenomena of galvanism they little thought that in the starting muscles of a dead frog's limb lay the germ of that which would one day bind the remotest corners of the earth together with telegraph wires. The discoverer of formic acid little dreamed at the moment that chloroform and its accompanying blessings lay hid in his discovery. Pasteur, Pouchet, Tyndall, and many others now see in the practical application of Professor Lister the benefits arising to mankind from purely scientific investigation, undertaken in fact with a view to elucidate questions of a very different nature. Professor Lister, of Edinburgh, combining in himself the rare qualification of the acuteness of the man of science, the skill of the experimenter, and the dexterity of the surgeon, has based on a scientific foundation a mode of practice which disarms of their dangers many of the worst injuries and the gravest operations.

K 2

LECTURE VIII.

THE PREVENTION OF ARTISANS' DISEASES.

By E. D. MAPOTHER, M.D.,

Professor of Physiology, Royal College of Surgeons, Medical Officer of Health, and
Surgeon to St. Vincent's Hospital, Dublin.

THE industrial classes of Dublin, according to the Census just
issued, number 50,943, or over one-fifth of our population,
and as they are usually the bread-winners of families, wide-
spread poverty must follow their loss of health. The cheer-
fulness which accompanies manual labour, with its ample
and regular reward, and the dignified feeling that the artisan
works for those at home whom he loves, are highly conducive to
health. There are, however, many employments at present
pursued under cruel conditions, which legislation and good
feeling on the part of the employers may remove or mitigate.
The richer classes should not forget that they owe their
comforts and luxuries to artisans, and that our country's
prosperity depends on their well-being, and that of the generations to succeed them. For such reasons I feel sure of
your sympathy, whilst I detail in a way, which must needs
be dry and fragmentary, some plans for the prevention of
the ills incidental to their labour. Workpeople will themselves most advantageously find out precautionary measures
as soon as, by the diffusion of the knowledge of physiology,
they learn to set a proper value upon health. The endeavour to teach them this knowledge is not new in Dublin,
for I find that 46 years ago a course of lectures on artisans'
health was delivered in the Mechanics' Hall by my earliest
professional friend, Dr. M'Keever.

The special diseases which ill regulated trades promote
may be arranged in three classes :—1. Those due to the
entrance of dust into the lungs. 2. Those due to slow
poisoning. 3. Those which constrained positions or overwork in close rooms engender.

I. The millions of little waving hairs which coat our air
passages resist dust for a long time, but increasing attacks
from without at last tire them. Steel-grinders suffer most
severely from the entrance of particles into their lungs.
The average duration of life at Sheffield among forkmakers,
who work exclusively at a dry stone, is but 29 years; those
who make scissors last 32 years, the rougher work being
done with a wet stone; and sawgrinders live on till 38, wet
stones being alone used. Just as desire for promotion makes
the soldier hail the battle, these men show a desperate

disregard for all precautions, and freely declare that if life were prolonged the employment would become overstocked, and wages consequently lowered. In Ireland wet grinding is alone employed, and that on a very small scale. As in the case of many other unhealthy occupations, safeguards should be enforced under the supervision of inspectors, or of the proprietors, who are often willing to do their duty in this respect. Magnetized wire respirators and magnets hung through the rooms effectually catch the iron dust, which is given off very largely; for instance, a razor loses half an ounce in being shaped from the rough. Professor Sigerson has figured these and other dust from various work-places, and I show you some of his diagrams. The eyes are sometimes protected by spectacles, and the need for this is shown by the fact that in a few months the glasses get opaque from particles imbedded while red hot. The grinders have to shape the stone rough from the quarry, and this should be done in the open air. A flue and fan would carry off the dust if each stone was boxed round, except at one point for working. One proprietor (Mr. Rogers) by such appliances has prolonged the lives of dry grinders to 46, and of wet grinders to 49 years. While the hair of the head is worn so long that it catches dust, that natural protective, the beard, is shorn by the workmen in this and many other dusty trades. Nature, by denying this dust filter to women, indicated that they should be exempted from labour at dusty occupations.

Stonecutters suffer from one-third more sickness than carpenters. This lung, taken from the body of one, contains stony particles and gray dust, which have set up slow inflammation. Instead of a pink, fleecy, and spongy mass, it has become black and tough, like Indian-rubber. The French stone with which the millstones of all countries are made, is, as you see by these specimens, so very hard and flinty that the chisel has to be sharpened every 20 minutes. It is largely worked in Dublin, and the men cannot stand the breathing of its chips more than eight years, although they suffer little for the first three or four. The very intelligent gentleman who conducts this business in Dublin has often to force his men to go to the more harmless occupation of cutting ordinary stone; but as they earn up to £3 weekly by piecework, they reluctantly consent, notwithstanding the danger to their lives. One London employer confesses to the killing of ninety men during the forty years he has carried on business.

Metal miners die by lung diseases in England in a proportion nine times greater than the agricultural population ; and as 1,075 deaths from accidents occurred among the 370,881 coal miners of Great Britain in 1871, it would appear that underground trades eclipse those pursued above ground as much in risk as they do in discomfort. Any of us who will cramp his body under a table some two feet high for half an hour, will understand what the collier may have to endure while for many hours he squats thus in the dark, and works hard with the pickaxe meanwhile. His day's labour over, he drags his body up the shaft, from a temperature beyond our summer heat to the darkness and chilliness of night, to descend again—at least during half the year—before daylight dawns next day. It is no wonder, therefore, that his lung is stuffed with dust, and burst from his efforts in breathing, or that his body is stiffened from such a posture and rheumatic chills. At the top of the shafts at the coal mines at Valenciennes, and those of Lady Bassett, in Cornwall, there are baths and comfortable rooms for change of clothes. They should be universal; and let us hope that if our mining operations should extend, inspection will be established and these appliances introduced in Ireland. Double shafts, or aspirators, and steam manlifts, should be in every mine insisted on ; the former have improved health and prevented explosions, and the latter have greatly lessened heart disease. There is an undue proportion of lung illness among the 421 coalheavers and the 101 sweeps of Dublin.

Pottery workers are subject, in a great degree, to lung disease, the more so since sōme technical improvements have rendered a stoppage of work during frost unnecessary. There is now need for greater sanitary vigilance. The dust rises mainly from the floor, and the workrooms should be swept, after proper sprinkling, at least every morning before the operatives begin. I am happy to say that our Irish pottery at Belleek, where over 600 hands are employed, excels in the healthfulness, as much as it does in the beauty of its work.

Textile fabrics of many kinds have their victims, the working of flax being the most hurtful. Dr. Greenhow reported to the Privy Council, that in one factory at Pateley Bridge, Yorkshire, 23 out of 27 who hackled flax were habitually asthmatical. Anyone going into the hackling or carding rooms of a flax factory coughs and sneezes violently. Machinery has almost entirely displaced hackling and wool combing by hand—the latter, formerly done in the worker's dwelling, was hurtful to every inmate. The machine boys, aged from 13 to 18 years, suffer much from dust, or "pouce," as it is called, in the Belfast flax mills; while those who do not begin work till they are adults, bear the dusty work very well. Among others, what is called "mill fever" arises, and as it has a peculiar rash, and occurs but once in life, it is not unlike the eruptive fevers. At paper works, the teasing of the shoddy, at hemp dressers (a large industry in Dublin), and at marine stores the picking of rags, create a most stifling and hurtful dust. This is also the case in feather stores, as the roguish rustics overweight their pluckings with lime and dust.

The remedies for dusty trades are palpable enough. 1. To filter the air by a respirator. This one, which I devised more than three years ago, was found very effectual; but as those who showed regard for health and life by wearing it were laughed at by their fellow-workmen, it has been abandoned. You see it consists of a wire gauze covering the mouth and nose, lined by a layer of cotton wool, ¼ inch thick.

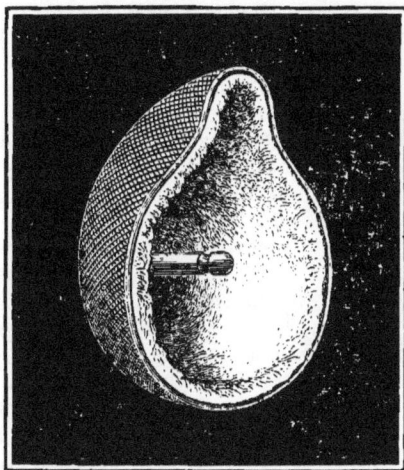

If the lining be thicker it flushes and heats the wearer. It is held by a piece of wood caught between the teeth, and thus the nose, our natural respirator, must be used for breathing. Mr. Pearson, of Ship-street, makes them for a few pence. If female workers wore bonnets or caps, with crape strings fixed across the mouth, much dust would be excluded. Dr. C. D. Purdon has invented a respirator, which Messrs. Grattan, of Belfast, sell for 13s. 6d. a dozen. It is made of buckram and cotton wadding, and is held on by two loops of elastic round the ears. Excellent results have followed its use in flax mills, potteries, and in Ward's great paper factory. 2. Ventilation by M'Kinnel's tubes or Archimedean tops, which I show you, or other special plans. Fine glass tubes are said to catch organic matter from the air passing through them, and a combination of such tubes might make a good ventilator between workrooms. 3. The action of steam fans, which in large factories have proved their efficacy by the increased appetite of all hands. Well arranged means for heating must be combined. In a Nottingham factory, heated by badly set hot air pipes, 190 out of the 200 hands were attacked with bronchitis. 4. The exclusion from all labour, requiring vigorous muscular and breathing efforts, of persons under 18, whose organs up to that are not tough enough to resist ill-usage. (Tyndall's experiments with dust in the air and sections of artisans' diseased lungs were then shown by the electric light.)

II. The second class includes those trades which lead to slow poisoning. Lead has long been the painters' bane; but since the introduction of zinc paints and the more general use of paper for the walls of parlours and halls, its ill effects are less frequently seen. Mr. Price, of the Midland Railway, finds that paint in which iron is substituted for white-lead is better for painting iron work. As the salts of lead are not volatile, it must be from the hands the poison is brought to the mouth, especially when food is being eaten. The metal gives a warning, for the gums "hoist a blue Peter" in the form of a lead coloured line along their edges. When this sign appears the workmen should have nothing to do with lead for a week or two. I believe this blue line, as well as the other coloured lines from copper, mercury, and other metals, to be due to the reduction of the metal by light, for it does not appear round the back teeth, which are kept in the dark. The sugar in the mouth helps the reduction. The other ill effects of the poison are great weakness from spoiling of the albuminous matters of the blood, and paralysis of the muscles on the back of the forearms, especially in painters and plumbers. This selection of the part is probably owing to the lead getting in through little breaks in the skin of the hands, as is proved by file makers, who fix the tool on a bed of lead, with the forefinger getting palsied there; and the potters who handle lead glaze suffering also in the forearms. Many painters are wholly disabled, and, consequently, pauperized. The muscle of the heart is amongst those which lead specially weakens. Great mortality used to follow the grinding of white-lead; but now that a moist process is substituted, little harm results, and lead-mining is the only occupation in which the poison actually kills. At Reeth, in Yorkshire, where half the men are lead-miners, the deaths by lung diseases are double those of the agricultural population. Women who work at lead seldom, if ever, bear children. The sweeping of the rooms, the floors being previously wetted very early in the morning, saves many workpeople. I am happy to say I find that the workpeople at Ballycorus lead works are most healthy, owing to perfect airiness, and the avoidance of cleansing flues on windy days, when dust would affect the men greatly. The preventives of lead poisoning are numerous and efficient:—1. Washing the hands and mouth frequently, and always before meals. I have known painters who have thus escaped during twenty years' work. Those who live far away from their work suffer most, as they take their meals without ablution. The addition of a few drops of sulphuric acid to the water in which the hands and mouth are washed would be a further preventive. 2. Ten drops of this acid to a pint of water makes a pleasant drink, which, by changing the lead into the most insoluble form, has checked poisoning in many lead works. Milk is also especially useful, by quickly renewing the albuminous matters in the blood which the metal spoils. Fermented liquors are remarkably injurious, as they lead to the fixing of the metal around the joints, and thus produce a kind of gout. 3. The wearing of a linen suit, which should be washed weekly. Last month a painter was in St. Vincent's Hospital for palsy of the lower limbs, which had resulted from his habit of washing the paint from his garments while they were on him. 4. The taking of a bath at least once a week. Metallic poisons tend to the skin especially, as is best shown in the case of silver. Five grains of nitrate of silver taken internally have dyed the whole skin gray, and the metal, therefore, cannot be equally distributed in the body. The Turkish bath, in which the skin is longest exposed to the action of light, would probably reduce the metal most effectually. 5. The lead colours should be mixed with the oil, or turpentine, in a machine something like a churn. 6. Painters in burning off old paint must inhale a good deal of lead. I have found that brushing over old paint with carbolic acid strips it thoroughly off, and the injurious scraping of the wood is avoided. 7. If painters with "dropped wrists" are mad enough to work on, an Indian-rubber strap from a glove on the hand to the coat above the elbow will give some help to the wasted muscles. It may be feared that painters and plumbers will not understand and value preventive measures, but I have always found them a particularly intelligent class, and of the 1,968 in Dublin, only 1-19th are unable to read and write. Printers suffer occasionally from lead, especially if type is delivered to them wet, when absorption rapidly occurs. If the "composing stick" is kept too full of type, weakness of the forearm muscles may follow. The antimony of the type appears to cause the profuse perspiration, slow pulse, and depression which they labour under occasionally. A few years ago the deaths among London printers between the ages of 35 and 55 were twice as numerous as among men of similar ages generally, but for this closeness of workrooms and overwork were to blame more than the effects of metals. On one Sunday newspaper the compositors had often to work, without a break, from Thursday morning to Saturday evening. The rooms were usually ventilated by windows which were rarely opened, as the draught made the gas-flame flicker, and as there was sure to be found among the men (as there so often is in a railway carriage) one so selfish and stupid as to insist on all the windows being shut. The workrooms of Dublin printers are better, but if they were more frequently whitewashed they would be more lightsome and cheerful. Roof ventilation should be universal. As in too many other trades the men are often idle for the earlier part of the week, and overworked for the latter part; masters should absolutely prevent this.

Copper among braziers and those who work with copper paints is the cause of great weakness, spitting of blood, and other distressing chest symptoms. A marone line round the gums, discovered by Sir D. Corrigan, forewarns of the danger. Some years ago I attended a fatal case of copper poisoning in St. Vincent's Hospital. The lad was wholly employed at scraping the old green paint from Venetian blinds and mixing the fresh colour. The pigment is Olympian green, or carbonate of copper, and a respirator and a mixing machine would render these processes harmless. As I have before stated, carbolic acid would strip the paint as fully as the use of sandpaper. Chrome green is a safe substitute for the copper colour. Braziers should never place the brass articles in their mouths, which they frequently do. Zinc rises in very dense fumes in brass foundries, and the workmen are often seized with an illness like ague, which obliges them to cease work for a few days. The casting should be done in the open air, or under well-arranged flues

Mercury is the most hurtful of all the metals, and miners, mirror makers, and gilders, who have to deal with it, lead lives inconceivably miserable. Constant salivation, low spirits, great weakness, and shaking palsy are almost universal; and cases have been recorded in which the limbs were so disabled that food had to be taken by the mouth as quadrupeds take it. Loss of mind has also resulted; and truly looking-glasses would reflect many a sad face if the sufferings of those who made them were remembered. Electricity has rendered gilding with the aid of mercury obsolete, and the making of mirrors from nitrate of silver, now so general in France, should abolish all the horrors of mercurial labour. You will see that this silvered mirror, which I got in Paris in January last, is as brilliant and perfect as could be; and moreover, it is cheaper and more lasting than the mercurial ones. The process of Petitjean, by which it was made, throws down the silver by ammonia and tartaric acid. The cost for silvering is only 1s. 8d. per square yard—one-third less than the deadly process by mercury. In England three processes have been invented —that of Drayton, using oil of cassia; that of Thomson and Mellish, which employs grape sugar; and that by which the specula of many telescopes are silvered, milk sugar being the reducing agent for the silver. The full details of the last three are given in "Cooley's Cyclopedia" (fifth edition), and I trust they may be studied and adopted by our mirror makers, whom I have found most anxious for means to render their trade less hurtful. If mercury is still to be used, the following are practicable measures for the mitigation of its evils:—1. Peculiarly liable persons should be excluded from work. Physicians know that a few grains of mercurial preparations will in some produce fierce effects, while others seem almost insusceptible of their action. 2. The work should be only allowed for two or three days in the week. 3. As the poison tends to the skin, baths are highly preventive, and the workmen of Negretti, the barometer maker, by a daily bath avoid all illness. 4. Flues and fans should be so arranged that all vapour and dust must be carried off. An idea of the quantity emitted in the workroom may be gained from the fact that one manufacturer recovered 20 lbs. of mercury by a single sweeping of his chimney. The dust would be also caught by double casing to the walls, the inner being pierced by holes. The mercury could be thus saved. 5. Milk and eggs should enter largely into the diet of the workers, for they neutralize the effects of the poison. Sulphur, taken internally, would also change the metal to the sulphide, the least active of its salts.   6. The use of the respirator previously described.   7. A linen suit should cover the whole body, except the face and hands.   8. At the factory at St. Gobain, sprinkling the floor and walls with ammonia is said to have been highly preventive.   At the government establishment in the Rue St. Denis I saw last January much disease, and few precautionary steps.

Arsenic in the form of Scheele's or emerald green, is the cause of great suffering to workpeople, to those who live in rooms coloured with it, and to those who breathe it in from certain ball dresses and wreaths.   I show you wall papers sold in Boston, London, and our own city, which contain this arsenite of copper largely.   Up to 70 grains per square foot, or 1 lb. of arsenic in the papering of an average room have been found in some specimens.   Such paper, when brushed over with ammonia, turns a bright blue, owing to the copper, and this is a readily applied and fairly reliable test, or the burning of a bit of it gives a strong garlic odour.   Artificial flower makers suffer severe and even fatal illnesses from this emerald green—painful rashes, sore eyes, sickness of stomach, and great weakness, being the symptoms.   The remedies are very clear—1. The sale of this poison should be restricted as white arsenic is, and all arsenical papers and articles of dresses should be marked "poisonous."   2. Work should be occasional—that is, other work should alternate with the dangerous one.   3. The work should be done at a table pierced with holes, through which the surplus powder would fall, and it might be collected in drawers below.   As dust diffuses itself so thoroughly, as you have seen, any antidotal substances for arsenic would meet it in the air.   4. The hands should be frequently washed with water containing one-tenth of hydrochloric acid.   The dangers from arsenical ball costumes any of us may encounter.   A wreath coloured with it contained 40 grains ; 3 oz. have been taken from a tarlatan dress by Professor Nichols, of Boston, and 60 grains were proved to have been thrown off from another dress in one evening's wear.   If many belles were dressed thus in a crowded ballroom there would be, indeed, a "dance of death." In the Emerald Isle it is a favourite colour, as one may best judge at a Patrick's ball.   It is to be hoped that some more natural and harmless tint of green will become fashionable, for thus alone will dangers from arsenical colours cease. Those who are loyal to the rival colour do not escape; for the common orange dye, chromate of lead, has lately caused numerous cases of poisoning.

Gas makers suffer in many towns from the noxious effluvia

given off by the refuse lime, and from excessive perspiration, unless gruel be used freely as a drink. The amount of perspiration given out by a workman in the retort-house is twenty times over the usual rate. In our gasworks the iron process has been, at my suggestion, adopted by the company, and most signal advantages have followed to the workmen and the citizens who live near. Matchmakers—or, to avoid an ambiguous term, I should rather say workers in phosphorus—number but 11 in Dublin; and all work in a well-conducted factory in Ardee-row. A small over-crowded dwelling-room was used for this purpose six years ago in Ball's-lane; but, after my evidence of injury to health, it was abandoned. Phosphorus in the English factories used to cause painful, and even fatal disease of the lower jaw; and that substance so saturated the workers that their clothes, for many days after leaving off work, were luminous in the dark. Preventive measures have been so far successful that, in the great Manchester factory, no disease has resulted since 1863, whereas, for the previous twenty years of its work, most distressing results were observed. The essential steps for prevention are—1. To forbid the dipping or mixing of the phosphorus compound in any room into which the other more numerous workers have access. 2. To forbid the taking of meals without washing the hands, or in any room, save that for cutting wood, as is prohibited by the Factory Act, 1864. 3. The exclusion from work of those who have bad teeth, for thus the poison reaches the jaw. 4. The wearing of a respirator, such as I have described, the cotton being wet with a solution of potash, to catch the fumes. 5. The hanging of cloths, soaked in turpentine, which would absorb the vaporized phosphorus. 6. And above all, the exclusive use of amorphous phosphorus, the red powder made by exposing common phosphorus to a heat of 400 degrees for some weeks. It does not vaporize, and is, therefore, harmless to workmen. Bryant and May use this; and further, they place it on the box, the match being tipped with a composition containing chlorate of potash. As the matches light only on the box, risk of fire is greatly lessened.

As regards the tobacco trade, cleanliness, ventilation, and the use of a respirator, would greatly lessen the weakness, functional heart diseases, and extraordinary complexions, which greatly trouble the workers. Their time of work should be short, with frequent intermissions. Rolland's torrefier is used in France, to the great advantage of the workpeople.

III. Those diseases which constrained positions or over-

work in close rooms engender. The baker suffers from the
entrance of dust into his lungs, but more from the circumstances just named. In Dublin in 1859 his work hours
averaged seventeen, beginning on Sunday afternoon about
four o'clock, and he slept usually on the empty sacks, or in
a bedroom over the bake-house, where the carbonic acid of
the air was excessive. For a wonder, Ireland was for once
before England in sanitary legislation, as regards the baker,
the 1 and 2 Vic., chap. 28, prohibiting any baking process
on Sunday, except the setting of Monday's sponge. Dr. Guy,
of King's College, London, described several of the bake-
houses in the metropolis as underground, without daylight,
pervaded by sulphurous and worse smells, often flooded, and
overrun with rats. One was not high enough for a man five
feet nine to stand erect in. It was no wonder that one-third
of the London bakers habitually spat blood. Good example
in Scotch towns, and in Belfast, and dignified agitation, procured the Bake-house Act, 1863, which provides for the
frequent cleaning of the premises, a wholesome sleeping-room,
and the exclusion from work, between 9 P.M. and 5 A.M., of
all lads under eighteen. Unfortunately the carrying out of
the Act is not left to the Factory Inspectors, and I fear it is
much neglected in some towns. The evils of the night-work
are still deplorable, and if the men took nights alternately
in setting the sponge, the other journeymen surely could be
spared between the afternoon and four the next morning.
Some years ago I found hours in flour mills even more ex-
acting, the dust more hurtful, and there was Sunday labour
in those worked by water. Last week one of our news-
papers had a letter, signed "A White Slave," which complained that in small confectionery shops the female bakers
had to work over thirteen hours daily, and in dark cellars
most usually.

Shoemakers nearly all work in their single dwelling-
room, and by the piece, the family often giving aid. They
inflict excessive hours on themselves—often from six in the
morning till midnight—alleging that they would be dis-
missed if work is not returned at a certain hour, however
unreasonable. Some of them idle every Monday or Tuesday.
Their stooping posture, with the work pressed against the
stomach, confinement in close rooms for so many hours, and
their poor diet, consisting largely of tea, render them the
most dyspeptic of all artisans. The circumstances of our
weavers are very similar.

The tailor's posture on the floor, with the legs crossed and
the head bent down, must be injurious, and apprentices

always complain of it. I feel sure the amount of apoplexy is excessive among the men in this trade; and giddiness on first rising up after long work is very common. A man sitting in this posture breathes with much less than half the vigour of a man walking at the rate of three miles an hour; and therefore we find that their deaths (many of which are by consumption) exceed those of agricultural labourers of similar ages by two-fifths. The breastbone of tailors is often bent greatly backwards. A six-inch seat and a foot high table, with semicircular notches at the edge, would afford a far more healthy posture for work, and would seem to the uninitiated just as handy. Nearly all the tailoring workrooms while the Workshop Act was carried out by the Corporation were found to be very fair, and in those which were not, amendment followed prosecution. Previously the hours were excessive, and midnight on Saturday was often encroached on. In 1864, Dr. E. Smith, of the English Local Government Board, reported as follows of sixteen of the most important West End tailoring shops:—"The cubic space in these ill-ventilated rooms allowed to each operative and the gaslight only 156 feet per man. . . . In another room, which can only be called a kennel, in a yard lighted from the roof and ventilated by a small skylight opening, five to six men work in a space of 112 cubic feet per man." The outwork system, which is often known by a term more sensational and less polite, "the sweating system," is in force with some of the second-rate tailors in Dublin. Its alleged advantages are that the man's wife and family may work and add to the earnings, and that exorbitant charges on the part of shop workers and strikes are prevented. Its evident evils are the withdrawal of the wife from household duty and of the children from school, the temptation to work at night and on Sundays, and the possible conveyance of contagious diseases from the families of the workpeople to those of the customers. It is further alleged that dissolute workmen, unable to get respectable employment, are taken into the rooms of families who are able to get work, which they thus sublet. I think you will agree with me that the evil outweighs all possible good.

Next we must consider the seamstress' case, and more briefly than the topic deserves. The Government Inspector in 1864 reported that in one London collar-making establishment, in which 400 females were employed, the proprietor confessed " that the handworkers could only get imperfect nutrition upon the low rate of wages (9s.), and said that they did not get meat more than occasionally, replacing it

by some highly flavoured substance which would enable
them to swallow the staple article of their food—dry bread.
The dinners of the girls, which he had noted with some
care, would consist often of a pennyworth of pickles and a
pennyworth of bread." What must be the lot of their
Dublin sisters, whose wages range from 3s. to 7s., to meet
prices now one-third higher? That low prices paid for
articles of ladies' dress account for this rate of wages few
husbands or fathers present will allow. Where, therefore,
does the money go to? While the Public Health Committee
carried out the Workshop Act there were many prosecutions
for overcrowding of dressmakers' rooms. In one case 22
girls were packed in a room which allowed but 136 cubic
feet space for each of them, or less than half the lowest
amount for healthy breathing. Weakness of sight, from
over use of the eyes with ill-arranged light, and indigestion
from bad and hasty meals and long sitting in a close room,
are diseases which every hospital and dispensary physician
has commonly observed amongst needle-workers, who number in Dublin 7,552. Such ills can be readily obviated by
having the light high above their heads, moderating daylight by pasting green tissue paper over the upper panes of
the windows, and giving variety in the colours of the
materials to be sewed. Regulation of work hours will be
presently considered. To the habit of giving orders immediately before great balls or shows, and to the increasing
dressiness of the period, deplorable overwork is often due.
Would that the fair sex understood that their admirers
would value their sympathy with the poor seamstress far
above splendour of attire. The sewing machine has surely
served the needle-worker, and no harm results if the work
be not continued for more than five hours unbroken. Toil
with the feet has led to illness from undue supply of blood
to the lower half of the body; but if all hands were taught
the machine, rest and work by turns could be allotted to
all. Steam and an ingenious magnetic apparatus have been
substituted for foot work at Salem, Massachusetts, and Hall's
and Parson's treadles (of which I show you figures) have
been advised for the lessening of this labour by the Board
of Health of that State.

Overwork of the fingers has also its ill consequences, as
these casts of the cramped state of scrivener's hands will show
you. The posture, too, with the spine twisted and the face bent
down, as if the nose were to be used as a pen, has injured
clerks and school pupils. A seat with a back to it and properly arranged light would set all these evils at naught, if

the hours were not unduly protracted. Too constant work with the letter stamp and with the hedge shears has produced a contraction somewhat similar to the cramp of scriveners. Constrained positions in the open air are more hurtful than the most muscular labour under the same circumstances, and thus the drivers of vehicles and watchmen suffer from bronchitis and rheumatism to a much greater degree than farm labourers, and consumption is readily promoted in those subject to it.

My hour is nearly exhausted, and, therefore, I must omit much which might be said about other trades, and end by enumerating the measures which are desirable, and, I trust, gradually attainable for the improvement of the health of our working-classes. 1. Reasonable hours of work, to be regulated by factory inspectors, of whom there are but two in Ireland. Enthusiastic and untiring as Mr. Astley is, his visits cannot be frequent enough over the factories and workshops of more than half our island. The Factory Acts loudly call for codification, or at least assimilation—for instance, in that regulating workshops there is no power to enforce the keeping of a registry of workpeople on which their ages and time of beginning and ending work can be seen, or the posting of the regulations of the workshop. Again, it is not certain if meal time is allowable to those who work less than twelve hours daily. A medical certificate of age, and, what is more reliable, of fitness for work, should be required from every workshop operative, as it is from every factory worker. I trust that shop attendants will share the advantages of such measures by the passing of Sir J. Lubbock's Shop Regulation Bill now before Parliament. During the forty years of their operation such statutes have conferred enormous benefits. The manager of a Bradford factory reports that forty-five years ago his father carried him on his back to work at five in the morning, and that, except half an hour for meals he laboured till eight at night. Then factory girls were so tired at night that they had not energy enough to take off their clothes when they reached their homes, and flat foot, in-knee, and curved spine were common among the youths of both sexes. 2. Occasional vacations in the more dusty and the lead and mercury occupations would do much to obviate their evils. A month in the year, a week in the quarter, or a half holiday in the middle of the week, would be the most useful periods for various trades, and in many cases agricultural or household work could very beneficially occupy the artisans meanwhile. If employment failed—an event inevitable in all trades—poverty would

L

not result if these occasional intervals had been well used. No full holiday is declared in the Workshop Act. In Dublin the Saturday half holiday has not been highly valued, but by permission of the inspector any other day may be substituted, and Wednesday being near the middle of the week, would be the most appropriate day. 3. Meals are partaken of by artisans very irregularly. If their dwellings happen to be near their place of work, it is better that the open air in going to and coming from dinner should be employed, and economy and social intercourse are usually promoted by the taking of this meal at home. Those dressmakers who live far away from their parents are often content with bread as a mid-day meal to support their toiling bodies. For such the provision of a dinner like that supplied to the assistants in all our great drapers' shops would be a most useful and not necessarily an extravagant step. If such refreshment rooms as that in South George's-street were multiplied the object would be achieved in another way. I must allow that the failure of the Workman's Hall in Kevin-street has discouraged many friends of the working classes.

The employment of married women in factories, who number in the United Kingdom a quarter of a million, and in workrooms, is for evident reasons deplorable, and at all events infants should not be deserted by their mothers till six months of age. Half the children born in the Staffordshire potteries die within the first two years, and over one-tenth of the deaths are due to burns and drowning. The fault mainly lies with the husbands, who are not manly enough to work hard and avoid the tavern, and thus keep their wives at home. Dr. Baker, Chief Factory Inspector, tells us of the wondrous blessings the Factory Acts have achieved, "without an atom of personal, commercial, or national wrong." If the time for labour has been shortened, the pay has advanced, our exportation of textile fabrics has enormously increased, and the operatives are more intellectual owing to the leisure afforded them. That moral safeguards are called for will appear from another extract from his last report :— "An intimate acquaintance with the working classes for 40 years and more (38 of which have been official)—their habits their language, the certainty of their early induction into vice from the admixtures of the sexes, the disregard of parental control by the young from the sudden independence of wages, and the means thus afforded of injurious indulgences enables me to speak on these topics with some authority, and only confirms what might have been anticipated from the employment of boys and girls and men and women together, without any moral supervision. To these causes may also be added, the one sleeping-room of the homes of so many of them, the adolescents washing and dressing in the presence of each other, the unsexing nature of so many of the employments, such as foundries, furnaces, and brickyards; and the drunkenness and bad examples of many of the parents, all of which have operated, and are still operating, to produce a lamentable state of immorality amongst them."

Baths free or very cheap are sadly needed in several of the more crowded parts of the city. Ablution is much neglected by artisans, and one sees them in the galleries of theatres as sooty and grimy as when they left work. Excursions and walks on the Sunday afternoon do much to preserve the artisan's health, and proximity of sea and mountain makes Dublin of all cities that in which they can be more readily enjoyed, especially as we have now the tramways. Too many of the labouring class spend the greater part of the Sabbath and the Monday idly in bed or in drinking. Employers should punish all those who absent themselves on Monday. Social economies, such as savings banks and Government assurances and annuities, which are so much safer than "Friendly Societies," do much for the mental quiet, and, therefore, bodily health of the working classes. The opening up of our squares would prove a great blessing to the children of the working classes. Again, there are many spaces now disused or covered with ruins, which, if levelled and planted with a few trees and grass, would make capital recreation grounds. For example, the Cabbage Garden, Long-lane ; the spaces between Patrick's Close and Bull-alley ; between Great Elbow-lane and Gill-square ; between the rere of the Adelaide Hospital and Golden-lane ; and on the north side, Newgate. The demolition of many tenements utterly beyond repair, and the substitution of decent dwellings for the industrial classes in the city or in the suburbs, connected with it by cheap tramways, are among the most pressing requirements of the day. Our taxation, which, for water supply and main drainage, has provided, or must provide three-quarters of a million for the twelve years (1864–75), can scarcely bear unusual expenditure for some years to come. Our city is, however, conspicuous for the generosity of its prince merchants, and benevolence could scarcely find more worthy objects than those I have hastily indicated.

L 2

LECTURE IX.

ON THE CONSTRUCTION OF DWELLINGS,

WITH A VIEW TO THEIR

SANITARY ARRANGEMENTS.

By GEORGE C. HENDERSON,
Architect.

THE construction of dwellings with regard to their sanitary condition is the subject which it has fallen to my lot to bring under your notice; and the questions involved have lately been so thoroughly discussed, whether in the public journals, or in the more strictly speaking architectural and engineering publications, that I feel it would be quite impossible for me to say anything new, unless, indeed, I were to propound some Quixotic, crude, and untried theories of my own; but it seems to me the purposes of this lecture will be best fulfilled by a review of some of the best known maxims, and by calling attention to the proper application of the modern patents and inventions of sanitary engineers, and the subject being one of grave importance I much wish that the task of elucidating it had devolved on abler and more experienced shoulders than mine.

Before going into any details, I would preface my remarks by saying, that though we are bound to follow in the march of civilization, and as we should each, whether by original invention or the application of improvements, endeavour to improve the sanitary condition of our neighbours, we likewise ought to use our discretion in these matters, and not allow ourselves to be carried away by each article that appears in a *popular magazine*, for as fault-finding is easy to the scientific mind, it is only when remedies are sought, that the first *difficulties of the new* and the *perfections of the old systems* will appear; and though we must thank the Prince of Wales's illness for many sanitary improvements, we must also remember that many of the suggestions on sanitary matters made since are only good on paper.

Again I must ask you to bear in mind, that people do not always practise what they preach, for this simple reason: that to carry the precept into effect generally costs more than it is the practice to expend in this country; for what at the present day is the ordinary history of a dwelling-house from the time of its first conception until its final accomplishment?

Mr. A goes to architect B. I require, he says, so many reception and bed rooms of such-and-such a size, besides servants' apartments and stabling, and my limit of expenditure is so-and-so; all to be done in best manner with latest improvements. Mr. B, the architect, suggests he does not think it can be done for the money. Oh, says A, "Sharp" got his house for £1,000 less; it is quite as large and very handsome. Well, when the tenders for the work come in they are too high, and then the reductions commence, one by one the improvements are lopped off; in fact, Mr. A must have the quart put into the pint bottle, and as he will not hear of the reception-rooms, or elevation being reduced, for fear they would be inferior to his friend Sharp's, the contract is closed for the building without any of the latest improvements.

What I wish to lay before you is what a house should be, irrespective of cost; and *any dwelling to be healthful must be well drained and dry; must have an ample supply of water, as without this, from want of flushing, the best drains will become defective; it must also be well ventilated and heated.*

In considering the present subject, the denomination *dwellings* must be divided into *three* heads:

1stly. The country seat.
2ndly. The town or suburban residence.
3rdly. The artisan's or operative's house.

The site for a country house should be chosen with great care. A fine view is doubtless a great inducement to select a certain spot, but many good houses have been rendered uncomfortable by being placed in exposed positions; the ground selected should itself be elevated, and if it be commanded by higher lands, the arterial drainage from them should be diverted so as not to pass near the proposed building; and the plantings or natural undulations of the ground, if there be any, should be made available as a shelter from the west and south-westerly winds, which blowing almost constantly in this country are so charged with damp that they penetrate the most substantial walling, and are, therefore, more to be dreaded by the builder than the easterly winds, prevalent only during the spring months.

The next consideration is the plan of the house; this must of course be in accordance with the requirements of the family; in all cases, however, attention must be paid to the position of the various apartments: for instance, breakfast-rooms should have the morning sun, which passing westward ought next to strike the drawing-room, &c., while in libraries or billiard-rooms, the eye prefers a strong light without the glare of the sun's rays; again nurseries and dressing-rooms require the sunlight in preference to ordinary bed-rooms, and all sculleries, lavatories, &c., should have direct light and ventilation, through the outer walls; and if there be any fall in the ground, the kitchen should be placed at the lowest point for reasons that will be hereafter explained, and it will be found a great advantage to have the kitchen and offices outside the house and only one story high. The plan being agreed upon, next comes the drainage or sewerage. Now I have often heard people condemn drains and call houses unhealthy because the sewage only discharged into a cesspool. Do such persons ever consider what they are saying, and do they remember that there are not corporations and main drainage schemes all over the country? Now suppose, as they will probably suggest, we discharge our sewers into the adjacent stream; well as this supplies our neighbour next below with water for domestic use he may not think it a neighbourly act, and having got, say, Dr. Cameron to analyze it, may go to law with you, under some of the recent Acts against the pollution of streams (the operation of which will, I hope, soon be extended to Ireland). Now how do we stand? We have no public main, we can't go into the river, our neighbour won't let us, and neither will science, which says we must save all the sewage for the farm; so what is to be done? Build a good tank, and decide upon its situation at once; this should be as far away from and as far below the water supply for house as circumstances will permit. Now the main sewer should be laid in, making sure that you have a good fall; pipes certainly make the best drain, and the best of them I have seen are known as Jennings' patent, as they admit of examination by simply removing one of the collars; the lower half of the collar should be laid in cement, and the upper half should be bedded in common brick clay, known by builders as "puddle." A soft water cistern should next be constructed under ground, close to, but on the lower side of the house.

Both these tanks should have overflows, and if these overflows can be run into the garden they will be found useful for the purposes of irrigation, &c.

The use of large sized sewer pipes is a great mistake. People reason that as a building is large so it requires large sewers. You will at once see the fallacy of this argument when you consider that the sewage does not lodge in the pipes, but only passes through them, and the smaller the pipe the better the flush; so, therefore, it is the soil-tank or receptacle that should be made proportionate to the number of the inmates and not the bore of the drains, and for ordinary establishments a 9-inch main with 6-inch branches will be found sufficient.

I believe the best system of sewerage to consist of *three* distinct lines of drain pipes : the *first* taking the soil from house necessaries, scullery, and stables, &c., direct to tank; the *second* will take the water from bath-rooms, wash-houses, &c., and should be connected with the overflow from soil-tank; the *third* row will connect with the stack-pipes and convey the rain-water to the soft water cistern. Each of these drains should, close to the tanks, be provided with syphon-traps, with stand-pipes carried to surface of ground, and protected by stone with perforated metal cover, hinged; this has a twofold use: 1st, if any gases are generated in the tanks they will escape here by the stand-pipe instead of entering the house; and 2ndly, should any stoppage occur in the drains it will most probably be at this point where it can easily be removed.

The water supply is a part of our subject on which little can be said in a lecture, as the circumstances vary so in each case; above all, however, this portion of the work and the arterial drainage should be left in the hands of a civil engineer with permission to consult an analyst as to the purity of the water. Practically the main house supply must always be derived from either wells or local rivers, and, if possible, should be taken from such a level that, with a ball-cock, it will fill the house cisterns; or if the water have to be taken from a low level, if a stream serve, it can be used to turn a turbine-wheel. Now as many people have a mistaken idea that a turbine, like the hydraulic-ram, of itself actually forces the water up, I may remark that it only produces the motive power to work pumps which can draw from a well of filtered water, whereas the ram is a pump that can only raise the same water it is worked by. All water, whether hard or soft, should be filtered before it enters the dwelling, both on sanitary grounds and to prevent, as far as possible, deposit in the cisterns and incrustations on the pipes. A filter bed can be formed at the end of any tank by placing divisions across it; these form compartments to be filled with various beds of coarse and fine sand and animal charcoal, through which the water percolates into the tank. These filter beds should be periodically cleaned out and renewed.

The main cistern should be placed at the highest possible point in the house, but still in an accessible position, and should be of large size, say containing sufficient water to meet two days' consumption should any accident occur to the supply pipes. The consumption of water in Dublin at present is about 40 gallons per head per day, but Mr. Neville hopes to reduce this to about 30 gallons; and, taking this as a fair average, a family of *six* persons will require a cistern-area, as nearly as possible 6 feet long, 4 feet wide, and 2 feet 6 inches deep, containing in round numbers 375 gallons, weighing, as nearly as possible, 33½ cwt. exclusive of the cisterns; so it will be seen that some substantial bearers are required for the main cisterns. From the main cistern all the minor supplies should be taken to the lower levels, by which means the small service cisterns will always remain full.

The foundations demand special care. When the trenches have been carefully opened, and the level applied to see that, though practically level, there is a slight fall to one point, then the masonry may be commenced in the ordinary way, building the first course of large stones laid crossways to the wall and without mortar; from this rubble work may be carried up to the finish level of ground; and here I may mention that basements, properly so called, on sanitary grounds are not desirable. Houses having basement stories are, I am aware, usually spoken of as particularly dry; too often, however, this has only reference to the family or upper portions of the house, while the kitchens and servants' apartments will unheeded exhibit ground-damp to a large extent. Now, I believe our great ancestors were *as* alive to the unpleasantnesses of a damp dwelling as we are, and that, therefore, they built vaults under and round their houses to secure dry lodgings over; and in the olden days, when wines were kept on draught, the beer and cider made at home, and falconry, bear hunting, and cock fighting, were popular, doubtless good uses could be made of the subterranean apartments as stores and kennels for the "pets." As time crept on, however, the old form of building was often followed, but I believe the original use of the basement was lost sight of, and probably from want of accommodation for the retainers on the upper floors, the underground story first became a human habitation, and so the practice has been handed down from generation to generation to the present time when the introduction of basement floors is on most sites to be discouraged: 1st, they tend to dwarf the building, and so lessen the effect of the elevation. 2ndly, the story under, or half under, ground costs quite as much as a similar amount of work above the surface. 3rdly, in carrying off the drainage from the building, the excavation will be so deep, and the cost of removing any obstruction that may occur, consequently so great, that I for one would consider this alone a fatal objection; besides all this however, modern manufactures properly applied enable us to dispense with the basement, for what I set out as *its first use*, namely, to prevent damp rising, from a naturally wet soil or badly drained foundations, into the walls of our habitations; for this purpose a course of some substance perfectly impervious to water should be laid on all walls, say three inches over the ground. The various materials used are : slates or flags bedded in cement, bricks laid in the same material, coal-tar, felt bedded in coal-tar, perforated terro-metallic ware, and asphalte. To *most* of these there are objections; slates and flags are so brittle that if not very carefully laid they break with the superincumbent weight, and so become useless; *bricks* will absorb about one-fifteenth of their own weight of water, that is to say, a solid red-brick when soaked will be about seven ounces heavier than when it came out of the kiln; and if any one side of a brick be kept wet the moisture will soak through it, and even communicate with anything that may be in contact with the other side, and constant wet, helped by frost, will eventually rot even the best qualities; we therefore see that the brick is of little use, and that the non-conducting agent in this case is the cement which is as brittle as the slate, and therefore open to the same objections. Coal-tar is a very desirable material, and the cost so trifling that it is to be regretted it is not more generally adopted; in large buildings however, where the weight is great, tar alone is liable to be displaced, amd hence *two* courses of felt, each bedded in coal-tar, are constantly used in this country with satisfactory results ; but as *felt* exposed to the weather rapidly becomes porous, I believe it alone would be useless as a dampcourse. The tiles are made of various thicknesses, and of widths to suit all walls, and being glazed are impervious, so that wet can only ascend through the joints; the tiles will bear great pressure, and it has been urged in their favour that as a means of ventilation they are good; that may be so for a storehouse, &c., but I am inclined to think such a volume of air under a floor as would pass through a course of these would be rather unmanageable.

I would much prefer asphalte, say one inch thick, to any other dampcourse, as it can be laid without any jointings, and is possessed of an elasticity that in a great measure does away with the risk of fracture. This course being finished, a drain from the trenches containing foundations must be constructed to carry off the surface water, or that accumulating from springs should such unfortunately appear ; for this purpose the common field drain-pipe, or an ordinary ridge tile, laid in loose stones with proper falls will be found efficacious ; this drain should discharge into a small reservoir, which can be readily formed by sinking an old barrel, which a fireclay pipe trapped by one of the syphons above mentioned, should connect with the overflows from soil and soft water tanks.

Having so far described the progress of a house we find at least a good system of sewerage, and a dry bed on which to raise the superstructure ; we will now watch the building more as the critic than the constructor, suggesting what ought, and what ought not to be done. We see the ground floor joists laid ; below them, and just above the damp-course, one or two pieces of perforated metal will be placed under each room ; this it is asserted is to give ventilation and so preserve the timbers—a laudable effort, doubtless ; but if these opens be exposed to a strong wind, will not the cold air passing through the floors, lower the temperature, and produce draughts in the rooms over ? Now, how is this to be avoided ? Well, before answering this question, I would ask you to remark that while I maintain all ventilation should, as far as possible, be self-acting, I also desire that it should be under complete control. To prevent the cold rising through the floors, and the carpets becoming like small balloons partially inflated, as I have seen in suburban houses, I recommend the floors to be double, or counter floored, the space between the boards being filled in with, if near the sea-side, dry sea-sand ; if inland, dry sawdust. Next, these ventilators should be made to open and close, and as iron would, by rust, soon become fixed, it seems to me that something of a like form, made of brass, copper, or zinc, would best serve the purpose.

Now it is from the spaces under the floors to which a constant current of air can be admitted, that I believe the main supply should be brought into the house by flues, carried up to the backs of the fireplaces in each room, where there should be a chamber formed in terra-cotta or fire-clay (air passing over hot metals becomes unhealthy) ; here the air will become warmed, and from thence it can be conveyed into the apartment at any convenient point in the chimney-stack, the opening by which it is introduced should be fitted with a regulating valve; for if too much air be admitted below, or if the fire in the grate be strong, the current passing into the room will require adjustment; and as these air-shafts terminate below the point where the ordinary smoke-flue from the fireplace commences, it will be seen that there must be space for them in any building. It may be said that this system will only work where firing is used. Not so; a very small rise in temperature will rarify air, and put it in motion, and I believe that if any smoke-flue in a stack of chimneys be in use, it will generate sufficient heat to work all the ventilating shafts leading to the rooms over. Again, this is but a small objection, for if in July or August you are warm, if the walls of a room were honeycombed with ventilators of various sorts, it would be hard to per-suade my hearers that any of them worked half so well as the simple, but of late much abused open window. In winter time the case is different, and the popular want is thorough ventilation, without cold draughts, *when* the temperature is low; and I submit for your consideration a well-known system that can be economically, and, I believe, effectually carried out in new houses.

In old houses a good and simple mode of introducing cold air into the apartment consists of having a deep slip along the sill of the sash on the inside, which allows the window to open say two inches from the bottom, and so admits the air into the apartment by the meeting rail.

Ventilation, however, cannot be good unless the foul air be carried off in about the same proportion as the fresh is supplied. The accomplishment of this is full of difficulties. We know that most of the means in use are defective, but who can point out the right course to adopt? We will examine carefully the ordinary appliances:—One of these, known as Arnott's or Sherrington's Patents, is to be found in almost every modern house, near the ceilings of the principal rooms, sometimes fixed in the exterior walls, with perforations on the outer face, sometimes fixed communicating with the chimney. Arnold's Patent is a nicely balanced valve, supposed to permit the exit of all foul air, &c., while the income draughts caused by wind close the ventilator, which can also be closed at will by a cord attached to a lever. They are undoubtedly good in principle, and the best of their kind I have seen : lately I introduced one into a chimney ; there was no direct return smoke, but the draught being bad, sufficient soot escaped to blacken the ceiling in a few weeks; again, when fixed in an outer wall, in spite of the cork-buffer, the noise is sometimes so great that

I have known them closed up altogether, for people will not use anything that annoys them. These Sherrington's are perfectly simple, but require constant attention, while Arnott's only require attention when objectionable, and are therefore greatly to be preferred.

Another common practice is to leave an opening over the chandeliers, in the centre of the room, where a box is formed from which a pipe is taken through the outer wall, and generally turned up; while the temperature of the room is high, the chandelier burning, and the outer air calm, the bad gases *will* escape through the tube, but if the wind be high a return draught is to be feared, and when the temperature in the apartment is low, and but a small fire burning, there will nearly always be a direct current from this air-shaft to the fireplace. As a proof of this, I may mention that the sunlight burner, now so much used in our public buildings, when unlit, has been found to produce such strong currents of air across the heads of an audience that they are now generally fitted with "Mercury Floats," patented by the Messrs. Strode, of London; their name almost explains the working; when the gas is unlit there is no passage for the income or exit of air; light the burner, and the heat by expansion raises the float, and permits of the escape of foul air and the products of combustion; other appliances having the same object might be noticed, but we must return to our subject. In passing I may, however, remark that the sunlight in operation is founded on exactly the same principle as M'Kennell's ventilator, explained in Dr. Mapother's lecture. Small flues are also often carried up in the chimney stacks, starting just below the ceiling level, and they generally discharge a little over the slating; in the main, I believe this to be a good system, but I would like to have the ventilators carried the full height of the chimney stack; if, however, the flue be terminated near the ridge of the building, there should be an opening left on both sides of the chimney, so that if the wind strike one side the air may escape by the other. In buildings where there are large assemblages, or in public asylums, this system can be developed by carrying a number of small flues into a main shaft in which a current of air is maintained by the use of fire heat, or even a jet of gas kept constantly burning. The objection urged against the use of these flue-ventilators is that if the temperature in an apartment without a fire be raised, say by gas, to any moderate degree, air will enter the room by the chimney, coming down from exactly the point the foul-air shafts discharge at, and therefore it is asserted some of the poisonous gases will be reintroduced into the house. With all respect for the writers who hold this theory, I think it a little far-fetched; for we may take it as the broad rule that it is the tendency of all bad or unwholesome air to ascend. Therefore if it once gets to the top of a chimney, I don't see why it should or how it could immediately rush down the next flue, and even supposing it possible, it would be greatly diluted by fresh air and so be harmless, and if there be a register (as there should be) to the grate, the escape into the room will be almost nil; and further, I believe that down draught in chimneys is mainly caused by an insufficient supply of air in the apartment, and if the cause be removed the objection will also vanish.

Another mode of ventilation consists of a panel hung on pivots over the doors, or of sliding perforated panels in the doors; these have been much discussed of late, and as they communicate invariably with the interior of the house, they bring us to consider whether passages or the stairs should or should not be used as ventilation shafts; for my part I think not, for as there are two currents in every doorway—through the lower half, air entering the apartment—through the upper half, air rushing out in an impure state, then ascending by the stairs—the supply on each floor would become more and more fœtid; we should not therefore turn more bad air into the stairs than we can help, and the air in passages should be kept pure by the same means as suggested for the rooms. Many more details connected with ventilation might be discussed with interest, but the last I will touch upon has reference to the plumber's work. Many are the theories as to the proper mode of connecting the interior drains with the exterior already described. Here again these syphons come to our aid, for in this position from a stand-pipe a few inches from the wall of our house we are able for the last time to ventilate our sewer; but in all cases let it be remembered that the lead or soil pipe should come through the walls to the outside; but as to whether soil-pipes should be fixed inside or outside, I believe it to be quite immaterial provided all the joints be made in solder, and that these pipes be provided with ventilating tubes which may be connected with the main shaft rising from the sewer which should be carried to discharge at the highest point of the building, if possible in some tourelle or gable, away from the chimneys carrying the ventilation flues from rooms. The various pans and basins should be fitted with S traps in preference to those known as the D form of trap, which I am glad to say is now almost out of

date. Penetrating a little further into that labyrinth of
pipes to be found in a house well fitted and supplied with
soft water, pump water for general use, and hot water, we
find under the bath-cocks, the cleaning screws in traps of
lavatories and butler's pantries, &c., little leaden trays
placed to receive any leakages that may occur, and also in
the cisterns we have what are called stand-pipes; they serve
as overflows, and should also screw out at the bottom, so as
to allow of the cisterns being thoroughly cleaned. All these
pipes are connected with some one of the drains and are
seldom trapped, and indeed trapping would be almost
useless, for unless filled with water a trap ceases to be a
trap ; and through many of these pipes, used only as precautions, water never runs, consequently they should not be
introduced into a soil-pipe; they may be connected with the
service-pipe of a water-closet, but the waste of a bath should
be chosen in preference, and it will be better still if they
are run into one of the rain-water pipes.

I believe the junction of these rain-water, or stack-pipes
as they are called, with the main sewers now alone requires
notice. If the separate drains, such as I have described, be
used, all that will be required is to let the water discharge
on the surface of the ground into a dished stone, in which
is fixed a metal trap connected with the drain-pipe ; these
traps, besides taking the surface water, catch the sand and
other deposits washed off the roof, and can easily be cleaned
out.

Heating I must only briefly touch upon; and if we ever
give up our open fireplaces and adopt stoves or other means
of heating I believe the time is so far distant that such a
contingency need not now be considered seriously. In
speaking of ventilation a means of heating the air entering
the apartments has been referred to in connexion with the
ordinary grates, and with a view to procuring a uniform
temperature through the dwelling, and a consequent reduction in the consumption of fuel, as well as a diminution in
the risk of colds from draughts, &c. It will be found that
the hall, or lowest point of the stairs, is in practice, the
position to which the greatest amount of heat should
be supplied, as from thence it will diffuse itself through the
other passages, and so into the various apartments. A large
supply of hot air might be derived from a chamber formed
behind the kitchen range, and distributed by flues through
the house, and if a stove be used for the hall the well-known
" Musgrave " still holds the highest position, followed perhaps by the " Gurney." Hot water, however, should be the

means of heating aimed at, and not given up without a struggle: 1st, because the warmth generated is more wholesome and also more pleasant to the senses than that produced in any other way I know of except the open grate.

2ndly. Because if the house be not very large and a close range be used for the cooking the heat can be obtained without an extra fire; and here comes in the reason (promised in the early part of this lecture) for putting the kitchen at a lower level than the rest of the house: it is to enable us to make use of the fire in warming the water for the general supply of the house, as well as for heating purposes, as the boiler for any circulating apparatus must be at a lower level than the pipes; as the water when heated flows out at the top of the boiler and returns cool to the lower portion, many people use the same boiler to supply the house and heat the coil of pipes placed, say in the hall; so far as my experience goes I have not found this work well, as the heat is seldom sufficiently strong; I have however used with success twin boilers, which in winter, when required for heating purposes, work independently, and in the summer season can, by stop-cocks, be both turned on to the bath supply.

The Genus gas-stove has doubtless been much improved upon lately, but they are still in their infancy, and very undesirable if used without flues.

If time permitted there are many items in a house that might be usefully discussed; but I must now leave this very imperfect notice of the country residence, and pass to the considerations in connexion with town and suburban residences; here the tide of fashion hardly leaves us the choice of selection in our site, as too often the speculative builder builds in a rising locality, and the public are content to pay him a high per-centage for his outlay. To the affluent about to build in or near a town I would say that most of my remarks already made regarding foundations, dampcourses, ventilation, and heating will equally apply to their case; but as the main sewers are in these positions generally found ready made, the question of drainage is much simplified. To persons of moderate means I would say, *do* not build unless you can do it in the best manner; if you cannot afford it this, or next, year, wait until you have more money saved; in the meantime, if you or indeed any of my audience think your rented house unhealthy or the drainage bad, do not vex or worry yourself until, through very fear, you become ill, but get some practical man to examine the house for you, when I venture to say that by the expenditure of a very few pounds, in the manner suggested by him, most probably in putting only a ventilating tube to the soil-pipe, you will have as healthy a dwelling as any cheap house built by some common builder to your own special order. If your house be cold, and there be any fireplaces in the inner walls, by resetting the grates, you can have a chamber formed at the back, which fitted with a simple sliding valve below and above, the heat will cause a draught, by which a supply of heated air will be supplied to the adjacent room or passage.

· Here I must remark that though our local domestic buildings have improved in comfort, we can still learn many lessons from an inspection of any of the new villa residences near the English and Scotch towns, where, for instance, you will find a bath-room, and hot and cold water laid to almost every apartment in houses having a rental of say about £100 a year (in Birkenhead I know much cheaper houses having a bath-room and water supply in two dressing-rooms); and all these trifles tending to a plentiful use of water, whether for ablutionary or flushing purposes, have a beneficial sanitary effect.

Turning to the dwellings for the middle classes, of which we have seen hundreds built on both sides of the city during the last few years, I must condemn those landlords who, purchasing waste lands, have them laid out in small holdings, let them at enormously reproductive rents, and having laid a main sewer down the proposed thoroughfare, are content to leave the roads without water-tables, and to be formed of any rubbish that may accumulate as the buildings progress, and while they insist on a house of a certain value being erected, sufficient to cover and secure their headrent, make no covenants for the exclusion of middens and large ashpits from the confined yards attached to such buildings; and when we come to look at the artisans' houses, we find the same state of things in an exaggerated form still before us in this city. I would notice what may be called the Church-road district, now under demolition for the Dublin and Drogheda Railway extension. Here there were numerous small avenues of cottages, all inhabited by workmen employed in the neighbourhood. They brought their owners large profits, and the headrents were high; the houses were badly built, there are no drains, and in only a very exceptional case is there any water laid to the dwelling; the roads were never made, and are to this day ankle-deep in mud, and impassable for any spring vehicle. I will make no further remarks on these premises, but ask you to say can habitations under such conditions be healthful ? I might multiply instances by describing the filth in the yards and the squalor in the dwellings I have had to visit occasionally, and where certainly the weekly tenants get bad value for their money. Much good work is being effected by the Corporation Sanitary Inspectors, and also by the constant jogging of their memories the officials now receive from the new Sanitary Association. Reform, however, in this direction must strike deep to the root of the evil, and I believe legislation should define the requirements of a tenement holding, viz. :—proper drainage, ample water supply, and the minimum cubical area requisite for the inhabitants, all which, together with the proper formation of the roads in the case of new buildings, the local authorities or special agents should have power to enforce in the same way as the late Factories Act.

Were this accomplished, our artisan and working population, alike jealous of infringement on their rights, and alive to every small act of kindness conferred upon them, would soon see that they and they alone were the gainers by the transaction, and would soon be glad to leave their present haunts for the improved accommodation, as a most nefarious system of tenement lettings at present exists, I mean in those cases where an almost pauper becomes the *person responsible to the landlord.* Here the only way of making up the rent is by the husband's uncertain earnings, and what the wife can *screw* out of the rooms, consisting of the worst accommodation, let at the highest weekly rents, and under such circumstances repairs, papering, paint, or even simple whitewash, are things unheard of, and to the localities in which they are situated *fresh* air never penetrates. It is such houses I desire to close ; *this* accomplished, the philanthropic and charitable should step in and meet the demand for rooms by supplying model dwellings at a small increased rent. I know that speculations of this class have not in this town yielded a good return for the money invested, but looking back you will see that first efforts seldom succeed, it takes time to educate the public mind—for example, the tramways, which signally failed with George Francis Train, are now realities before our eyes ; so with the recent increase in wages, the demand for comfortable houses has also increased among the respectable operative class.

In this country, however, each family has an ambition to occupy a separate house, so that I fear such buildings as those started by the Industrial Tenements Company in this town, or by " Peabody," in London, will not ever be popular. Cottages containing say *three* apartments, of very moderate

M

dimensions, built round a square, each having a scullery or wash-house attached, besides a small yard for laundry purposes, good drains, water laid on, and the heat from the kitchen or living-room fire utilized, would seem to me the most desirable class of building. In the angles of the square I would place common latrines and ash-pits, while the centre would be occupied by a general superintendent's and collector's house, and a club room for the men.

The right to inspect all the premises, at reasonable hours, should form part of the tenant's agreement, and if the services of a retired sergeant of police, or some person of that class were secured as resident superintendent, I feel confident the managers would have but little trouble, the sanitary condition of the inmates would be good, and the shareholders would have a fair interest for their outlay, as well as the satisfactory moral reflection of having done some good in that walk of life to which it hath pleased God to call them.

LECTURE X.

ON SANITARY LEGISLATION.

BY ROBERT O'BRIEN FURLONG, ESQ., M.A.,
Barrister-at-Law.

LADIES AND GENTLEMEN,—The subject upon which I have to address you—" Sanitary Legislation "—is, I fear, by no means an attractive one. A moment's consideration, however, will suffice to convince you that it is a subject of the very highest practical importance to the sanitarian.

During the present course of lectures our attention has been directed to a variety of very interesting topics. Chemistry, meteorology, geography, medicine, and architecture, considered in their relation to the great subject of public health, have all in turn been brought under our notice ; and it was considered that the series would be incomplete if it did not include a lecture on the legal aspects of the question—for experience has shown that sanitary science cannot make its way in the world without the assistance of the Legislature.

To illustrate my meaning, I may refer to one of the earlier lectures of this course, in which we were told how to discriminate between pure and adulterated foods.

Now, all the information which our friend, Dr. Reynolds, could give us on this most important subject, would not prevent our being poisoned by dishonest traders, if there were not an Act of Parliament imposing a penalty on persons selling adulterated food, and providing for the appointment of public analysts to detect adulteration.

Again—to advert to another lecture—it will not suffice to tell the public that overcrowding, will probably result in an outbreak of " typhus." The overcrowding will take place in spite of the knowledge, and the fever will probably follow. The sanitary legislator must come to Dr. Grimshaw's assistance with stringent regulations, and an efficient organization for securing their observance, and then we may expect some diminution of this form of " preventable disease."

Sanitary legislation deals with the laws and regulations for the improvement of the public health, whether in the form of Acts of Parliament, or of rules and regulations made by the various local authorities throughout the country.

M 2

It is obvious that the conditions "which science and experience inform us are necessary for the preservation of health are not to be secured without the intervention of the state."

Individuals may do their utmost to live in obedience to the natural laws of health; but in the majority of instances, they will find their efforts frustrated by the neglect of their neighbours. For one man who respects these laws multitudes live in open defiance of them.

Hence arises the *necessity* for sanitary legislation.

The *general principle* upon which it proceeds has been thus stated by Mr. Simon (the distinguished medical officer of the Local Government Board):—

"All such states of property, and all such modes of personal action, *or inaction*, as may be of danger to the public health, must be brought within the scope of summary procedure and prevention."

The same idea is expressed by Dr. Acland* in more popular language. He says:—

"The true principle of sanitary legislation is that the Government should help the people to do, *not* what they *can* do, but what they *cannot* do. It should strive to ascertain what hindrances there are in the way of the people's health, and to remove those they cannot remove themselves.

"If the people like to live in dirty houses let them, so long as thereby they do not affect their neighbours; but let it not be impossible for them to have clean houses.

"The laws should aim at securing to all whatever is *necessary* to health. Nothing is too minute for the attention of the Legislature provided the minute point is *essential.*"

The subject of sanitary administration, is so closely connected with that of sanitary legislation, that I think it deserves some share of our attention; indeed a lecture on the laws relating to public health would be of very little practical use, if it did not contain some reference to the machinery provided by the state, and by the local authorities, for carrying out those laws.

My subject is so very extensive that it would be quite useless for me to attempt to treat it in anything like an exhaustive manner. I have therefore resolved to content myself with giving you a short sketch of the progress of sanitary legislation; I shall then make some general observations on the law as it stands at present and the machinery for carrying it into effect.

* Address at meeting of Social Science Association, 1872.

First, then—as to the progress of sanitary legislation :—
The first law relating to public health was passed so long
ago as the year 1388.* It imposes a fine of no less a sum than
£20, upon "all who cast annoyances, garbages, entrails, &c., in
ditches and rivers." The mayors and bailiffs of the cities
and towns were charged with the execution of this Act. It
is interesting to observe that, even at this early period, the
local governing bodies were intrusted with the care of the
public health.

About a hundred years later we find an Act† to prohibit the slaughtering of cattle in cities, and boroughs, "lest
sickness might be engendered unto the destruction of the
people."

The Sanitary Commissioners (1869) tell us that the
local courts were in possession of sanitary powers, and they
mention an instance of the exercise of such powers, which
is furnished by the Court Rolls of Stratford-on-Avon. It
appears that, in 1552, Shakespeare's father was fined for
depositing rubbish in the public street of that town, in
violation of the bye-laws of the manor ; and again, in 1557,
he and five other gentlemen were " amerced in the sum of
four pence each for not keeping their gutters clean."‡

Dr. Haughton called attention in his lecture, to the fact
that great epidemics always produce wonderful activity
about sanitary matters.

This remark applies with special force to our sanitary laws,
which we owe for the most part to the panics occasioned
by outbreaks of pestilence.

An attack of the plague in the reign of James I. led to
the passing of an Act§ which made it a capital felony for any
person suffering from the disease to go about, when com-
manded by the proper authority to keep his house.||

Again, in the years 1817–19, a fearful epidemic of fever
raged in Ireland, during which about a million and a half
cases (of which 65,000 proved fatal) occurred, and the fever
hospitals in Dublin alone treated 42,000 patients,—*i.e.* more
than one-sixth of the population.¶

* 12 Richard II., cap. 13. † 4 Hen. VII., cap. 3.
Both these Acts were repealed by 19 & 20 Vic., cap. 64.
‡ The same year Mr. Shakespeare became a member of the Corporation of
Stratford, and subsequently was appointed one of four "affeerors," whose duty it
was to impose fines on their fellow-townsmen for offences against the bye-laws of
the borough ; and so he ceased to be the subject of any further prosecutions.
Unlike the nuisance authorities of our time, these "affeerors," we are told, "dis-
played unusual vigilance, and considerable severity, " in enforcing the laws.—
Collier's " Shakespeare," vol. i., p. lxvi.
§ 1 Jac. I., cap. 31.
|| None of the foregoing Acts applied to Ireland.
¶ Barker & Cheyne's Hist. Sketch, vol. i., pp. 62, 94.

In 1818 a select committee of the House of Commons was
appointed to report on the progress of the fever, and to suggest measures to arrest its further extension.   The same year
an Act was passed for Ireland,* enabling the Lord Lieutenant
on the appearance of fever or other contagious distemper in
any town, to appoint a Board of Health, with powers to do
everything necessary for preventing the spread of contagion,
and restoring the sick to health.   The Board might cause
the streets to be cleansed; houses, yards, &c., to be purified;
beds and bedding to be disinfected, or burned; and all
nuisances prejudicial to health to be removed.   The funds
necessary for carrying out this Act were to be advanced by
the Lord Lieutenant, and repaid by presentments on the county.
A penalty not exceeding £5 was imposed on persons resisting
the orders of the Board, and default in payment was punish-
able with imprisonment for three months, without appeal.

In 1819 another select committee of the House of Commons
was appointed to inquire into the state of disease in Ireland;
as to the working of the Act of the previous year; and
the causes of the prevalence of fever amongst the working
classes.   This committee, in their report, recommended the
establishment of a systematic local control in all cities and
towns for the removal of nuisances which generate and
promote disease; and to this end they suggested the appointment of officers of health in cities and towns, to be elected
by the householders, and to be armed with ample powers
for the removal of nuisances, and the preservation of health.
In the same year an Act was passed which embodied the
principal recommendation of this committee.† Dr. Mapother‡
mentions in proof of the high estimation in which this Act was
held that " An appeal that its operations should be extended
to England was made by the famous Dr. Paris and Mr.
Fonblanque," but without success.§

This Act provided for the annual election by the vestry
of not less than two, or more than five, officers of health, in
every city or town containing more than 1,000 inhabitants,
or elsewhere, as the Lord Lieutenant should direct.

These officers of health were to act without salary, and
the expenses incurred by them in the execution of their
duties were to be levied as other parochial assessments.

The powers given to these officers, were somewhat more

* 58 Geo. III., cap. 47.   † 59 Geo. III., cap. 41.
‡ Statistical and Social Inquiry Society of Ireland.   Journal, vol. iv., p. 203.
§ Dr. Mapother also draws attention to the fact that the *first* Parliamentary
Reports on Public Health related to Ireland, and that public attention was *first*
called to the dangers of intramural sepulture, by the disgraceful state of a Dublin
cemetery—Bully's-acre.

extensive than those conferred by the former Act on Boards of Health.

Both these Acts were repealed by the "Sanitary Act, 1866."*

In the year 1836 a most important step was taken in Sanitary Legislation, by the passing of an Act for the Registration of Births and Deaths, in England and Wales. This system of registration was extended to Scotland in 1854, and to Ireland not until 1864.

It seems strange that Parliament should so long have withheld from Ireland the advantages of a system which had been established in England some twenty-eight years before, and which had long been in operation among all the other nations of Europe.

The Sanitary Commissioners (1869), observe :—

"That recent sanitary legislation has been remarkably drawn out by, and connected with, three outbreaks of cholera, which led to investigations of the means of preventing infectious diseases, and so drew attention to the fact, that the seats of endemic disease are generally where the air or water is polluted."

The first visitation of cholera occurred in the year 1831, and by a temporary Act passed in the following year,† which, strange to say, was not made applicable to Ireland, the powers of the Privy Council, under the Quarantine Act, 1825,‡ were greatly extended, and the Council was enabled to issue such orders as might be deemed expedient for preventing the spread of the disease ; "for the relief of persons affected thereby ; or for the interment of those who fell victims to its ravages."

By an Act of the following year,§ these powers were continued until the next session of Parliament, by which time the cholera had disappeared, and the Act was accordingly allowed to expire. The possibility of a future visitation of the disease does not appear to have been contemplated by the Legislature. This is a very good sample of short-sighted and unstatesmanlike legislation.

In 1840 an Act was passed for England, Wales, and Ireland‖ which empowered Boards of Guardians and overseers to arrange for the gratuitous vaccination of all persons resident in their unions and parishes. This Act has since been repealed¶ and another substituted for it.**

* In Thom's Directory for the present year (1873), page 9, the latter of these Acts is referred to as being still in force.
† 2 and 3 Wm. IV., cap. 10.
‡ 6 Geo. IV., cap. 78. The earliest quarantine laws are to be found in an Edict of Justinian, A.D. 542.   § 3 and 4 Wm. IV., cap. 75.
‖ 3 and 4 Vic., cap. 29. See also 4 and 5 Vic., cap. 32.
¶ By 30 and 31 Vic., cap. 84.   ** 26 and 27 Vic., cap. 52.

To give some idea of the beneficial results of this legislation, I may mention, on the authority of Dr. Haughton, that
the death-rate from small-pox, which before the introduction
of vaccination was 66 per cent. of persons attacked, has
now been reduced to 6 or 7 per cent.

In 1842 a most important series of inquiries as to the
state of public health was set on foot by the English Poor
Law Board. The first of the series was conducted by Mr.
Chadwick, then Secretary to the Board. The results of his
investigations are contained in a very valuable report, which
did much to arouse public attention to the necessity for improved sanitary legislation.*

In 1843 a Royal Commission was appointed under the
presidency of the Duke of Buccleuch, to investigate the
causes of prevalent disease and the best means of improving
the public health. This Commission made two reports.

It is much to be regretted that the admirable recommendations contained in their Second Report were not forthwith
carried into effect by a comprehensive Sanitary Act.

In 1846 provision was made for the removal of nuisances
and prevention of diseases, by a temporary Act,† which
applied to the United Kingdom.

Two years later a law was passed, "which may be regarded
as the groundwork of our sanitary legislation."

"The Public Health Act, 1848,"‡ unlike many of those
to which I have already referred, was intended to be permanent. It was framed with great care, and it is of a
most comprehensive character. Unfortunately it applied
to England and Wales only. However, many of its provisions have since been extended to Ireland, by subsequent
statutes.

This Act was of a permissive character. It was to be in
force only in any locality in England or Wales, (exclusive of
the metropolis) the ratepayers of which presented a petition for its application, or in which the death-rate was
exceptionally high.

It provided for the appointment of a General Board of
Health, to consist of three members, and to be continued
for five years.

The town councils in corporate towns, and in other places,
local boards, to be elected by owners and ratepayers, were
constituted the local authorities for carrying out the Act.

* I have omitted all mention of the numerous Acts relating to the government
of towns, which, although they contain important sanitary provisions, can scarcely
be regarded as "Sanitary Acts."
† 9 and 10 Vic., cap. 96.   ‡ 11 and 12 Vic., cap. 63.

In the same year the Nuisance Removal and Diseases Prevention Act of 1846, being about to expire, a similar Act was substituted for it,* and in 1849† this Act of 1848 was amended.

These two last mentioned Acts remained in force for Ireland until the year 1866, when they were repealed by the Sanitary Act of that year.

The third visitation of cholera, which occurred in 1854, led to some important changes in the sanitary laws.

The Nuisance Removal Acts of 1848 and 1849, were repealed for England, and a comprehensive Act‡ was substituted for them, which, however, did not apply to Ireland.

The powers vested in the Privy Council by the former Acts of 1846 and 1848, to make regulations for the prevention of formidable contagious diseases, were largely extended, and the provisions relating to them were embodied in a separate statute.§

In 1858 the General Board of Health, which had been created in 1848, and reconstructed in 1854, was allowed to expire, and its powers were vested in the Privy Council.

In the same year was passed the " Local Government Act, 1858,"‖ for England and Wales, which is to be construed with the " Public Health Act, 1848," as one Act. Several of its provisions have been extended to Ireland by subsequent enactments.

The new Act greatly extended local powers for the execution of sanitary works in such districts as adopted it, and gave in fact "most of the requisite powers of police and municipal government, if only they were duly sought, and duly used."

The " Nuisance Removal and Diseases Prevention Act, 1860,"¶ amended the two Acts of 1855, and amongst other things, reconstituted the Boards of Guardians as the " Local Authorities" for executing the Diseases Prevention Act.

In 1865 an Act** was passed, which applied to Ireland, enabling local authorities to dispose of the sewage of their respective districts, in order to prevent its becoming a nuisance ; and to make arrangements for the application of such sewage to land, for agricultural purposes.

The local bodies were constituted " Sewer Authorities," with powers to construct sewers; to take lands compulsorily, and to prevent the pollution of streams. "This Act may be

---

* 11 and 12 Vic., cap. 123.
† 12 and 13 Vic., cap. 3.
‡ 18 and 19 Vic., cap. 21.
§ 18 and 19 Vic., cap. 116.

‖ 21 and 22 Vic., cap. 98.
¶ 23 and 24 Vic., cap. 77.
** 28 and 29 Vic., cap. 75.

said to have introduced into rural districts the first real instalment of active sanitary powers."

In the year 1865, the Town Council of this city presented a memorial to the Chief Secretary, calling his attention to the defective state of the sanitary laws for Ireland; and to the action then taken by that body and their distinguished medical officer of health, I believe we owe the "Sanitary Act, 1866."

The history of this Act affords a curious illustration of the hasty and unsatisfactory manner, in which our sanitary code has been formed.

At the time to which I am now referring, the cholera was supposed to be advancing towards our shores, and the special provisions of the "Diseases Prevention Act, 1855," had been put in force in England.

This Act, as I have observed, did not then apply to Ireland, and "the Nuisance Removal and Diseases Prevention Acts" of 1848–49, which are declared in the preamble to the Act of 1855, to be "defective, so far as the same relate to the prevention, or mitigation of endemic, epidemic, or contagious diseases," afforded the only protection provided for us by law, against the approach or spread of the disease.

Notwithstanding this declaration as to their inefficiency, these Acts were considered good enough for Ireland, for ten years after they had been condemned in England.

The necessity for providing an immediate remedy for this anomalous state of things, was felt so strongly by the then government, that a bill was prepared, extending to Ireland in one statute the English statutes up to 1865. However, before this bill was brought into Parliament, it was ascertained that a complete sanitary code for England was in preparation, and accordingly the Irish Bill was altered, so as to include the latest provisions of the proposed English code.*

Towards the end of the session, it was found that there was not sufficient time to allow of the English Bill being passed through Parliament. Instead, therefore, of a complete and comprehensive statute, a fragmentary and somewhat confused bill was introduced, and passed for the United Kingdom; originally intended to apply to England only, but afterwards adapted so as to suit some of the more urgent requirements of Ireland also.

However although the form of the legislation was not all that could have been desired, the benefits conferred by the Act of 1866 were large and substantial.

* Dr. Hancock's Report on the " Sanitary Act, 1866," p. xvi.

I shall notice very briefly some of its provisions.

(1.) The authorities constituted by the Act were as follows :—

    I. In cities and towns corporate—"the Corporation."*

    II. In towns and townships—the "Town"—"Township"—"Lighting and Cleansing," or "Municipal" Commissioners, as the case may be.†

    III. In such part of each union as is not under another sewer or nuisance authority—"the Poor Law Guardians."

(2.) The obsolete Nuisance Acts of 1848-49 were repealed, and the Acts of 1855, with numerous alterations, and extended powers, were substituted for them.

(3.) The vestry officers of health were abolished, and *one* nuisance authority constituted for each district.

(4.) The expenses of the nuisance authorities were charged on the borough rate in corporate towns; town rate in towns under Commissioners, and the poor rate in districts under Boards of Guardians.

(5.) All nuisance authorities were *required* to make periodic inspections of their districts, with a view to ascertain what nuisances need abatement; and were empowered to appoint and pay Nuisance Inspectors.

(6.) The definition of a nuisance was extended so as to comprise

    I. Premises in a state
    II. Any pool, ditch, or water-course so foul as to be
    III. Any animal so kept as to be
    IV. Any accumulation of matter

      injurious to health.

These nuisances were included in the Act of 1855. The following were added to the list by the Act of 1866 :—

    V. Over-crowded houses.
    VI.   „   workshops.
    VII. Smoke from factory chimneys.

* The sanitary arrangements of Dublin are under the control of a Public Health Committee, consisting of about thirty members—all corporators. This committee may at any time be dissolved by the Corporation, or its members increased or diminished.

† According as the town is governed :—

1. By Commissioners under the "Towns Improvement (Ireland) Act, 1854," 17 & 18 Vic., cap. 113, or a local Act;

2. By Lighting and Cleansing Commissioners, under the 9th Geo. IV., cap. 82; or,

3. By Municipal Commissioners, under the 3 & 4 Vic., cap. 108.

(7.) Powers were given to the Privy Council in *England* to
extend the quarantine law to *coasting* vessels, in case of the
outbreak of contagious disease on board.   These laws had
previously applied to vessels coming from *foreign* ports
only.

(8.) The 49th section empowers the Lord Lieutenant,*
upon complaint being made that any *sewer authority*
has neglected to provide sufficient sewer accommodation ;
or to provide a supply of water, where the existing supply
is a source of danger to the inhabitants ; or that a *nuisance
authority* has made default in enforcing the provisions
of the Nuisance Removal Acts: to institute inquiry, and
if satisfied that the authority has been in default, to
limit a time for the performance of its duty in the matter
of such complaint, and if the duty be not performed within
the time so limited, to appoint some person to perform it,—
the expenses and costs to be recovered from the local authority.   Unfortunately this power is so little known that it is
scarcely, if ever, exercised.

(9.) Whenever any part of Ireland *appears* to be threatened
with, or affected by, any formidable epidemic, endemic, or
contagious disease, the Local Government Board *may* put
into force for a period of six months (which period may be
subsequently extended) the provisions of the "Diseases
Prevention Act, 1855," as amended by the "Nuisances
Removal and Diseases Prevention Act, 1860." The Boards
of Guardians are the "local authorities" for executing this
special Act, and all expenses incurred under it are to be
defrayed out of the poor rate.   In this city, which is under
the control of two Boards of Guardians—those of the North
and South Dublin Unions—the Local Government Board,
in the event of this Act being put in force, would probably exercise a power conferred on them by section 40,
and require the two boards to act together for the purposes
of the Act ; or they might direct that the town council
should be the "local authority" instead of the Board of
Guardians.   In the latter case the expenses of executing
the Act would fall on the borough fund instead of the
poor rates.   So long as the provisions of this Act remain
in force the Local Government Board may issue regulations—

1st. For the speedy interment of the dead.
2nd. For house to house visitation.

* This power has since been transferred to the Local Government Board
(Ireland).

3rd. For dispensing medicines, guarding against the spread of disease, and affording to persons afflicted by, or threatened with, such epidemic, endemic, or contagious diseases, such medical aid, and such accommodation as may be required.

The "local authority" is charged with the execution of the regulations so made, and has the power of appointing medical, and other officers, and of providing everything necessary to mitigate disease. This Act was not put in force during the recent visitation of small-pox,—although it can scarcely be denied that the disease assumed the character of a "*formidable* epidemic."

Two Amendment Acts were passed in 1868 and 1869, which were originally applicable to England only, but both have since been extended to Ireland.*

The latter of these Acts enables the Commissioners of Public Works to advance money required for sanitary works executed by the Local Government Board, in case of default of the local authority, the sums so advanced to be charged on the local rates.

Another Amendment Act† was passed in 1870 for the United Kingdom, but its provisions appear to me to be inapplicable to this country.‡

"The Artisans' and Labourers' Dwellings Act, 1868,"§ supplies a want which had long been complained of by local authorities.

In any place in which this Act is in force (*i.e.* in Ireland, towns corporate, boroughs, and towns under Commissioners), on the report of the local officer of health, that any premises within his district are in a condition dangerous to health, so as to be unfit for human habitation, the local authority (*i.e.* the Corporation or Town Commissioners), shall refer the matter to a surveyor to report on the cause of the evil complained of, and the remedy therefor, and whether the premises can be improved by structural alterations, or require demolition. On receiving his report, the authority shall require the owner of the premises to execute the necessary

---

* 31 & 32 Vic., c. 115, and 32 & 33 Vic., c. 100, extended by Local Government (Ireland) Act, 1871.

† 33 & 34 Vic., cap. 53.

‡ The latest specimen of sanitary legislation is "The Sanitary Act, 1866 (Ireland), Amendment Act, 1873" (36 & 37 Vic., cap. 78), which provides for the payment of the expenses incurred by a Port Sanitary Authority, under the Act of 1866, sec. 30, out of a common fund to be contributed by the riparian nuisance districts, (*i.e.* districts, any portion of which abuts on any river or coast forming part of a port), in such proportions as the Local Government Board shall order.

§ 31 and 32 Vict., cap. 130.

works, and in case of default, shall order the premises to be shut up or demolished ; or the authority may itself execute the works. (Secs. 5, 6, 18.) If the works be executed by the authority, such authority shall obtain from the Court of Quarter Sessions an order charging on the premises all expenses incurred, together with interest at the rate of £4 per cent. (Sec. 19.) If the owner execute the works he may obtain an order charging on the premises in which the improvements have been effected, an annuity at the rate of £6 per cent. on the amount expended—such annuity to be payable for thirty years to the owner and his representatives. (Sec. 25.)

The local authority may levy a special rate not exceeding two pence in the pound for any year, or may borrow money from the Commissioners of Public Works, to defray all expenses under the Act. (Secs. 31, 32.)

Any four or more householders living in or near to any street may represent to the officer of health that any premises in or near to that street, are unfit for human habitation, and the officer shall thereupon inspect, and report on such premises. (Sec. 12.) If the local authority neglect for three months after receiving such report to put the Act in force, the householders may address a memorial to the Local Government Board, asking for an inquiry, and upon receipt of such memorial the Board may direct the local authority to proceed under the provisions of the Act, and such direction shall be binding on the authority. (Sec. 13.)

The Act expressly provides that the absence of any representation by householders, shall not excuse the officer of health from inspecting any premises and reporting thereon. (Sec. 12.)

In 1869 a Royal Commission was appointed* to inquire into and report on the operation of the sanitary laws in England and Wales (except the metropolis), so far as these laws relate to sewerage, drainage, water supply, removal of refuse, control of buildings, prevention of overcrowding, and other means of promoting the public health ; and also to inquire into and report on the operation of the laws for the prevention of contagious diseases and epidemics, and the administration of all these laws, including the constitution of authorities, and the formation of areas.

For some unaccountable reason Ireland and Scotland were

* This Commission was appointed in compliance with a request submitted in May, 1868, to Her Majesty's Ministers, by a deputation from the Social Science and British Medical Associations.

not included in the terms of the Commission. The two countries, however, were worthily represented on the Commission, by Dr. Stokes and Sir Robert Christison.

The Commissioners examined a large number of witnesses, and presented a most valuable report, the general purport of which is thus briefly summed up :—

"The present fragmentary and confused sanitary legislation should be consolidated, and the administration of sanitary law should be made uniform, universal, and *imperative* throughout the kingdom."

Appended to the report is a summary of the existing sanitary law, together with suggestions for its amendment and consolidation.

In July, 1871, Sir C. Adderley, the Chairman of the Royal Commissioners, introduced a Bill into Parliament to consolidate and amend the laws relating to public health and local government; which was in fact the report of the Sanitary Commissioners in the form of a Bill. "The first part of the Bill proposed to repeal all existing sanitary laws, and of the 450 clauses of which it consisted, about nine-tenths were inserted, merely for the purpose of re-enactment."

The second part divided the whole country into sanitary districts, so that there should be no place without a sanitary authority, and only one such authority in each place. It was late in the session when this Bill was brought in, and it was read only a first time. So far as I am aware no attempt has been made during the present session to re-introduce it.

A very important Act was passed in the year 1871, which although not strictly a sanitary Act, so seriously affects the powers of local bodies that it deserves special mention.

"The Local Government (Ireland) Act, 1871,"* applies to towns in Ireland under any form of local government, whether the governing body consists of a Town Council, or Commissioners, under any general, or local Act.

The main object of the Act was to afford to local Boards an easy and inexpensive method of obtaining certain powers which they might desire to possess, and which until the passing of this statute could not be obtained without a special Act of Parliament,—a proceeding which, as the citizens of Dublin know too well, always entailed vast

* 34 & 35 Vict., cap. 109.

expense on the ratepayers.* Under this Act, if a local
Board seek powers to acquire land for public purposes;
to extend or reduce the limits of the district under their
control; to transfer from the Grand Jury to the local
Board their jurisdiction with regard to roads, bridges, and
public works; to sanction the making of further rates in
addition to the maximum authorized rates; to provide for
the future execution, or repeal, or alteration of any local
and personal Act in force within the town, they have only
(after some preliminaries have been complied with) to present a petition to the Local Government Board (Ireland),
who will thereupon direct an inquiry in respect of the
matters mentioned in the petition, and if the result of the
inquiry prove satisfactory, the desired powers will be conferred by a Provisional Order. This Order must be confirmed by Act of Parliament, which will not involve any
expense to the parties seeking the Order.

The Bill confirming the Order may of course be opposed
in its progress through either House; but in practice I think
it will be found that opposition will rarely be offered.

This Act also provides for the audit by an officer of the
Local Government Board, of all municipal accounts—a matter
of great importance, as one of the chief objections to sanitary
improvements is the increased expenditure they would entail
on the already over-taxed ratepayers. Possibly it may be
found that under the supervision of the Government auditor
a larger portion of the public funds may be made available
for sanitary purposes than is the case at present.†

" The Local Government Board (Ireland) Act, 1872," next
demands our attention.‡

Its object was to concentrate in one department the supervision of the laws relating to the public health, local
government, and the relief of the poor in Ireland.

* From a return lately presented to the House of Commons, it appears that a
sum of £18,199 19s. 2d., was spent by the Corporation of Dublin between the
years 1864 and 1872 in promoting, watching, or opposing the progress of Bills
in Parliament.
The costs of the Dublin Main Drainage Act of 1871 amounted to no less a sum
than £4,290!
† The following is an extract from the report of G. W. Finlay, esq.; the auditor
appointed by the Local Government Board to audit the accounts of the Dublin
Corporation for the year 1872:—"The amount of fines imposed under the sanitary
Acts but not levied is considerable. The Corporation are entitled to these fines,
and the Public Health Committee should see that the officers intrusted with the
duty of levying them, discharge it with promptitude, as the non-levying is likely to
have an injurious effect. It would be almost better that the Corporation should
not incur the cost and trouble of legal proceedings to enforce compliance with the
Acts, than, having done so, to allow the order of the magistrates to remain a dead
letter by not enforcing the penalties inflicted by them."—*Report*, p. 12.
‡ 35 & 36 Vict., cap. 69.

A similar Board for England, was formed in 1871, and is now presided over by a Cabinet Minister.*

The Irish Board consists of a president, ex-officio, being the Chief Secretary to the Lord Lieutenant for the time being, a vice-president, and two other members, one of whom must be a physician or surgeon, of not less than ten years' standing.

Amongst other improvements effected by this Act I may mention the following :—

1. The Poor Law Commission was abolished, and the functions of that body vested in the new Board.

2. All powers and duties vested in or imposed on the Lord Lieutenant, the Privy Council in Ireland, or the Chief Secretary to the Lord Lieutenant, by several Acts of Parliament, including the "Sanitary Act, 1866," and the statutes amending it; the "Nuisances Removal and Diseases Prevention Acts, 1855 and 1860," and "The Sewage Utilization Act, 1865," were transferred to, and imposed upon, the new Board.

The "Public Health Act, 1872," remains to be noticed. This Act owes its existence to a feeling on the part of Her Majesty's Government, that the then session of Parliament should not be allowed to close, until *some* improvement had been effected in our system of sanitary administration.

The Bill was introduced at such a late period of the session, that it was evident that any attempt to deal with so vast a subject in a large or comprehensive manner, must of necessity be futile. There was really no time for the discussion of details, and accordingly all the clauses likely to provoke hostile criticism, and so to imperil the fate of the Bill, were dropped, and a very short, partial, and incomplete measure was hurried through both Houses of Parliament, at the very close of the session.†

The Act applies to England and Wales only ; but it is pretty well understood that, if the pressure of public business permit, Parliament will be asked to sanction a similar measure for Ireland during the present session.

Time will not allow me to do more than notice very briefly, two of the leading provisions of this Act ;—

1. It divides the whole of England, with the exception of

* The Right Hon. James Stansfield, M.P. The other members of the Board are the Lord President, the five principal Secretaries of State, the Lord Privy Seal, and the Chancellor of the Exchequer. The constitution of the Irish Board is preferable, as it includes three paid members, who have to devote their entire time to the duties of the Board, and one of these *must* be a medical man. The want of a medical man on the English Board is to some extent supplied by the existence of a medical sub-department under the direction of Mr. Simon, F.R.S., who was for many years the chief medical officer of the Privy Council.

† The Lords bestowed only three days on its consideration.

the metropolis, into sanitary districts. These are of two kinds—*urban* and *rural.*

The urban districts are the boroughs, Improvement Act districts, and Local Government districts ; and the sanitary jurisdiction is vested in the Town Council, Improvement Act Commissioners, or Local Board, as the case may be.

Rural districts consist of such parts of the Poor Law unions, as are not included in urban districts; and the Guardians of each union form the rural sanitary authority.

The great advantage of this arrangement is, of course, that it is thoroughly exhaustive. No place in England or Wales is now without its local sanitary authority.

The Local Government Board is the central authority, and possesses the powers for sanitary purposes which formerly belonged to the Privy Council, the Home Office, the Poor Law Board, and other departments of the state.

2. By section 10, every Sanitary authority is *obliged* to appoint a medical officer of health, who in rural districts *may* be the medical officer of the union.

I have read many criticisms of this Act; and I think it is plain that it is not regarded with favour by the leading sanitarians of the sister country.

Dr. Rumsey, who is one of the first authorities in England on all matters relating to public health, while acknowledging that the promoters of the Act were actuated by an earnest desire, to propose what they believed to be best calculated to further the ultimate objects which sanitary reformers have in view, observes that—

"There are not a few matters of grave importance, in respect of which the Act has created obstacles more or less barring future progress, and very similar to those which have so greatly complicated and enhanced the difficulties of our present legislation."

Again, the "Joint Committee on State Medicine" of the British Medical and Social Science Associations* has condemned the Act as "entirely ignoring the cardinal principles, which they consider *essential* to any well-considered reform of the laws relating to public health, and of the authorities by whom such laws should be administered."

I admit that Dr. Acland seems to regard the Act with satisfaction, and of course his opinion is of the highest authority; but even he does not consider it as in any sense a complete measure.

Lord Napier and Ettrick, the President of the Social Sci-

* This "Joint Committee" was originally appointed at the meeting of the National Association for the promotion of Social Science, held in Dublin in 1861.

ence Association,* though endeavouring to say a good word
for the Act, is obliged to admit that it has no character of
finality. " If," he says, " we examine the powers of administration and action committed to the sanitary authorities,
we are at once involved in obscurity and confusion. The
Bill has given machinery, but it has not given faculties of
operation sufficiently categorical, distinct, and extensive."

I have not time to quote any other opinions, but I think
I have said enough to show that *some* at least of the most
eminent authorities in England regard the Act with dis-
favour. The leading medical journals condemn it also.

In my humble judgment it would be better, instead of
accepting this Act, *mutatis mutandis*, for Ireland, to wait a
little longer for a measure, or rather a series of measures—
for the entire subject is too large to be dealt with in one
Act—with some appearance of completeness.

No measure can be at all complete which does not supply
us with a simple and intelligible body of laws, in lieu of the
confused mass of statutes which at present compose our sanitary code. Everyone admits that one of the chief difficulties
in the way of effective sanitary administration, is due to the
large number of these statutes, and the way in which they
are framed.

These statutes are of two kinds—General and Local.

In not a few instances the Local Acts contain provisions
more or less at variance with the General Acts; and sometimes it happens that the ample powers of a General Act,
are obliged to give place to the more limited powers conferred by the Local Act.

Besides these Local Acts, there are bye-laws without
number, made under powers contained in various Acts, and
which themselves have the force of law.

The Sanitary Commissioners state that *these* are sometimes in conflict with the general law of the land.

The result is that it is often very difficult, and sometimes
utterly impossible, to ascertain the exact state of the law on
certain points.

In Dublin we have to search for our sanitary law amongst
a large variety of Local Acts, bye-laws, and regulations,
besides about seventeen General Acts. How, I would ask,
can the law be effectively administered when its provisions
are scattered over some fifty or sixty statutes, some self-
contradictory, others obsolete, and all of them vague and
loosely worded ?

* Inaugural address, Plymouth, 1872.

N 2

As the Sanitary Commissioners have well expressed it—

" The number of the statutes, and the way in which they have been framed, render the state of the sanitary laws unusually complex.

" This perplexity has arisen from the progressive and experimental character of modern sanitary legislation, which has led to the constant enlargement and extension of existing Acts, without any attempt at reconstruction, or any regard to arrangement. The fatal result is that the law is frequently unknown, and even when studied difficult to be understood."

Mr. Simon bears similar testimony. He says :—

" The laws which ought to be in the utmost possible degree simple, coherent, and intelligible, are often in nearly the utmost possible degree complex, disjointed, and obscure. Authorities and persons wishing to give them effect may often find almost insuperable difficulties in their way ; and authorities and persons *with contrary dispositions* can scarcely fail to find excuse or impunity, for any amount of malfeasance or *evasion.*"

It is much to be regretted that our legislative system provides no machinery for securing that an Act of Parliament shall be expressed in proper, and self-consistent language.

The great object of the promoters of a Bill is that it should pass ; and consequently they take care that it shall be framed, not in the most logical and appropriate language, but in that which is likely to excite the least opposition. Accordingly the draftsman is often obliged to choose "the path of obscurity and confusion, to repeal merely such scraps and fragments of the existing Acts as it is absolutely necessary to get rid of; perhaps also incorporating the old Acts with the new, thereby not only making the new Act unintelligible, except by reference to the former Acts, but raising a swarm of difficulties as to its construction, and its relation to the pre-existing law." This system of incorporation is one of the greatest obstacles in the way of sanitary action.

Sanitary Law is in fact a thing of shreds and patches. An eminent medical man has described it as being made up of "little dabs of doctoring done by different departments of Government." But Government is not to bear all the blame of the present state of the Sanitary Laws. Government has not originated the host of local Acts to which I have referred. *They* have sprung up somewhat after this fashion.

* Report, p. 21.

Some active member of a local authority discovers what he considers to be a defect in the law. Perhaps his fervid imagination has suggested to him some particular form of nuisance, which might not come within the definition of a nuisance contained in the existing Acts. The evil must be met by a new Act. His colleagues are persuaded that they are utterly helpless so long as this supposed defect is allowed to continue. Accordingly a Bill is promoted in Parliament at great expense. A number of heterogeneous provisions are crammed into it; another "Improvement Act—Amendment Act"—is added to our local code, and the mess is rendered more hopeless than ever.*

Traces of this mania for special legislation are to be found in some general Acts also.

A few years ago it became notorious that people were being poisoned in large numbers by means of arsenic. Immediately an Act of Parliament was passed to control the sale of this drug alone, as if there were no other poison known.

On this principle we should have a special Act to prohibit the sale of chloroform, another of strychnine, and so on, for every poison an Act. The absurdity reached its climax in the year 1853, when one of these intelligent legislators introduced a Bill for the prevention of glanders (a disease which killed one person in the United Kingdom last year).

Again, it sometimes happens that some provisions of the Sanitary laws are at variance with the general law of the land.

A curious instance of this is afforded by what is now well known as " The Birmingham Sewage difficulty." The case was this. For some time the sewage of the town was discharged into the river Tame, in strict accordance with the provisions of the Public Health Act, 1848. At length Sir Charles Adderley obtained an injunction from the Court of Chancery, restraining the Corporation from polluting the river. The Corporation then fell back on the provisions of another Act, and turned their sewage upon the land adjoining the sewer outfall. However, the landowners in the neighbourhood very naturally objected to this proceeding, and a second injunction was obtained. Here was a dilemma! The Corporation were compelled by the Act of Parliament to receive the sewage, but the Court of Chancery refused to let them dispose of it. At length the Corporation had recourse to a Bill to enable them to utilize the sewage on some waste lands far from the town. The Bill was opposed, and

* The necessity for these Local Acts has been greatly obviated by the passing of the Local Government (Ireland) Act, 1871. See *supra.*

thrown out, and, so far as I know, "the difficulty" still remains unsolved.*

Again, the law recognises a distinction between various Sanitary duties which seems to be based upon no intelligible principle.

To quote from the Report of the Sanitary Commissioners:—

"The removal of nuisances seems to have been considered a work apart from Local Government, and the prevention of epidemic disease, a work unconnected with the suppression of its causes. Such distinctions without differences, the result of *casual legislation*, regardless of what had preceded, are far more than merely illogical or unmeaning. They cause grave misunderstanding of the law, mislead public opinion, multiply expenses, and aggravate disinclination to improvement, and distrust of science."

The distinctions between authorities are equally unmeaning.

The local bodies charged with the execution of the "Sewage Utilization Acts," and the Sanitary Act, 1866, Part 1, are designated "*Sewer* Authorities." In the Nuisance Removal Acts, the same bodies are styled "*Nuisance* Authorities," while a third title is created by the "Diseases Prevention Act, 1855," viz., "*Local Authority*."

In towns such as Dublin, the nuisance authority is the Town Council; but the *local* authority under the Diseases Prevention Acts is the Board of Guardians. Thus, if it should unfortunately become necessary to put these latter Acts in force in this city, the Guardians of the North and South Dublin Unions would be the "*Local* Authorities" for enforcing their provisions, while the ordinary Sanitary functions would continue to be discharged by the Corporation. We might thus have three authorities all hard at work, each trying to do the other's business, and, as a necessary consequence, neglecting its own. It is true, as I have already observed, that the Local Government Board *might* substitute the Corporation for the Boards of Guardians; but, on the other hand, they are under no obligation to do so. Dr. Hancock states that the reason why the guardians are named as the authorities under the Diseases Prevention Acts is, that they have an organization of Medical Officers and hospital accommodation, and they also control the dispensary arrangements.

Before the passing of the "Sanitary Act, 1866," there were three distinct sets of Sanitary officers in some places, viz.,

* Since this lecture was delivered I have been informed by the Town Clerk of Birmingham that the Corporation are erecting sewage works under the direction of Mr. Hawksley, c.e. In both the Chancery suits further time for the completion of the works has been obtained.

those appointed by the parish, by the town authority, and by the guardians. So things are not so bad as they used to be.

There is one matter to which I have already referred, on which all Sanitary reformers seem to be agreed, viz., that the law should be consolidated, or rather that a new and complete code should be substituted for the existing *general* Acts.

I fear that the present Government does not intend to deal with the matter, for I find that an elaborate digest of the statutes relating to *urban* Sanitary authorities in England and Wales, has just been presented to Parliament in pursuance of a promise given last spring by Mr. Stansfield, when a question was raised with respect to the consolidation of the Sanitary Laws.*

Many suggestions have been made for the amendment of the Sanitary Laws. I cannot do more than enumerate a few of the most important.

1. Many of the provisions of the Sanitary Acts which are at present permissive, should be made compulsory on the local authorities.

2. The procedure for the compulsory abatement of nuisances should be simplified.

I am told that in this city, when a magistrate's order for the abatement of a nuisance is resisted, 14 or 15 visits of inspection must be made, and about 10 different forms filled up.

3. When an owner or occupier makes default, or cannot be made amenable, the local authority itself, should be compelled to execute the necessary works.

4. The provisions of the "Diseases Prevention Acts," with certain necessary modifications should be made permanent.

5. Provision should be made for the registration of sickness.

6. The law for the registration of deaths, which is manifestly imperfect, requires amendment. It might be well to provide that no interment should take place unless on production of a certificate of registration. This is the law in England.†

* In a note prefixed to this digest, the Secretary of the Local Government Board states that "The work must not be regarded as an authoritative interpretation of the law;" and that "no attempt has been made to construe the several enactments which *appear* to be in force, or to interpret any *doubtful* or *contradictory* passages contained in them."

† Deaths are sometimes not registered for months after they occur. The deaths registered in the Dublin district during the week ending November 15, 1873, were at the rate of 35 per 1,000 of the population. In London for the same week the rate was 25 per 1,000. The excess was caused by the registration of 45 deaths which had occurred in the South Dublin Union Workhouse during the *three* preceding weeks! The death-rate in fact affords no reliable test of the health of the population. Four deaths from small-pox registered in the week ending November 22, 1873, occurred in *January*, 1872!

7. More adequate provision should be made for the payment of expenses incurred under the various Acts.

8. Compulsory powers of acquiring land for sanitary purposes should be given to the local authorities.

But although the law is confused, and in some respects defective, still I am far from thinking that the unsatisfactory sanitary condition of the country at large, is due altogether to the state of the law.

Dr. Budd of Clifton, a very eminent sanitarian, tells us that the health officer of Bristol, Mr. Davies—

"Finds that with the exception of what relates to the establishment of Fever Hospitals, the Act of 1866 gives him all the powers he could desire for preventing the spread of contagious disease." It is in the highest degree re-assuring, adds Dr. Budd, " to hear from the same practised authority, that he scarcely ever finds, on the part of the people who are the subjects of them, any difficulty in enforcing the provisions of this Act."

Dr. Burke of the Registrar-General's department gave similar testimony before the Sanitary Commissioners. He says:—

" If the Act of 1866 be thoroughly carried out, there is very little more wanted."

I think it will be found that generally speaking the blame attaches not so much to the law, as to the persons charged with its administration.

I shall endeavour to show that this is so.

There are in Ireland about 274 sanitary authorities—163 rural, and about 111 urban.

The proposed " Public Health Act " would retain them all, with power to the Local Government Board to require some of them to combine, in certain cases.

It seems a mistake to have so many authorities.

In the single county of Cork (omitting the city) there are 10 urban, and 17 rural authorities; or one to about every 17,000 of the population.

Few will be found to deny that some, at least of the urban authorities, might be got rid of with advantage.

The Vice-President of the Local Government Board, Ireland, Sir Alfred Power, K.C.B., in his evidence before the late Sanitary Commission, gives it as his opinion that—

" The sanitary functions at present *supposed* to be discharged by the authorities in small towns, should be transferred to the Boards of Guardians, and the ordinary duties of local government be left in the hands of the Town Commissioners."

The evidence of the late Dr. Hill, who for many years occupied the important position of Medical Poor Law Inspector, is to the same effect.

He says that—

" Where guardians are the local authority, the sanitary Acts have been carried out very fairly, but where a Town Council, or Town Commissioners form the authority, they do pretty nearly as they like."

Dr. Hill attributes this to the fact that—

" Town authorities are debarred from doing their duties by the expense, and there is no one to compel them to do their duty but the Lord Lieutenant, while guardians generally adopt the recommendations of the Poor Law Commissioners."

The only way in which I can form an estimate as to the manner in which our numerous urban authorities at present discharge their duties, is by observing the sums they have expended on sanitary objects. The statistics supplied by Dr. Hancock in his return of Local Taxation in Ireland for the year 1871* afford a reliable test of the action taken by these authorities. I will quote a few of Dr. Hancock's figures.

I.—As to towns under the " Towns Improvement Act, 1854." These are 75 in number.

In 43 of these towns nothing at all appears to have been spent on sanitary purposes, *i.e.*, under the " Nuisance Removal Acts," or for making drains, or sewers, or other sanitary objects.

In 17 towns the collective expenditure was under £10.

The total expenditure for the whole 75 towns was £1,567.

In Munster (exclusive of Queenstown, where about £200 were spent) the entire sanitary expenditure reached the large sum of £79; and in Connaught it was £15.

In Parsonstown, an important town, with nearly 5,000 inhabitants, the total amount charged for sanitary purposes is four shillings. I confess I am curious to know what sanitary object absorbed this sum!

II.—In 14 towns under " Lighting and Cleansing Commissioners,"† the total amount charged is £85.

III.—In 11 towns under special Acts, including 7 townships in the county Dublin, the total expenditure was £1,392.

I can see no object in maintaining a larger number of local authorities than is absolutely necessary, and it appears to me that in many instances the sanitary functions of the urban authority might with advantage be transferred to

* The latest " Return "—that for 1871—was published only in February last.
† Constituted under 9 Geo., IV., cap. 82.

the Board of Guardians of the union in which the town is situated.*

If this course were adopted in the case of towns with less than 5,000 inhabitants, the number of urban authorities would be reduced from 117 to about 44. If 10,000 were taken as the standard the number would be further reduced to 15.

So much for the towns under Commissioners.

IV.—Let us now take a glance at the state of affairs in the corporate towns. There are ten of them† exclusive of the metropolis, containing together a population of more than 409,000 inhabitants.

Here surely we might expect to find some tokens of a vigorous sanitary administration.

Nothing of the sort.

In four towns nothing at all appears to have been spent, and the total amount disbursed by the ten corporate authorities was only £2,390.

So far it is plain that the authorities I have referred to have not availed themselves very largely of their powers.

In Dublin, where there is certainly more sanitary activity than elsewhere in Ireland, the total amount expended by the Public Health Committee for 1871 was £2,050; nearly £1,000 less than was spent on the fire brigade.§

That the defects in the sanitary condition of this city are not altogether, or even in great measure due to the want of legal powers by the Corporation is, I think, very clear to anyone who will take the trouble of inquiring what powers that body already possesses.

I am happy to be able to quote in support of this view, the opinion of our present medical Officer of Health, himself an eminent sanitarian, and in everyway qualified to pronounce on such a subject:—

Dr. Mapother referring‖ to one of the numerous local Acts to which I have before alluded, " The Dublin Improvement Acts Amendment Act, 1864," observes that:—

" So ample are the provisions of this Act that it has left scarcely anything to be desired in the way of sanitary legislation, and it

* Boards of Guardians possess one great advantage as compared with urban authorities. They include, as ex-officio members, owners of property, and men of good social position, who are not dependent for their seats on the votes of the very persons whom they are sometimes obliged to prosecute for breaches of the sanitary laws.

† Viz., Drogheda, Kilkenny, Wexford, Clonmel, Cork, Limerick, Waterford, Belfast, Londonderry, and Sligo.

‡ Drogheda, Clonmel, Waterford, and Londonderry.

§ The expense of the fire brigade for 1871 was £3,032 13s. 11d.

‖ Journal of the Statistical and Social Inquiry Society of Ireland, vol. iv., p. 250.

will be the fault of the Corporation and its officers if the most substantial benefits do not soon follow. The poorer classes of the city about 100,000 in number dwell in some 8,000 houses, each room usually let as a separate tenement, and the state of these dwellings *has been* deplorable in the extreme.

"The Corporation gained by the recent Act the power to compel the owners of these houses to put in thorough repair, and keep so, the roofs, walls, and chimneys; to have their windows kept clean and glazed, and movable *at top* and bottom; to keep a properly trapped house-drain and other sanitary requisites in good order."

Dr. Mapother further expresses his belief—

"That the sanitary state of Dublin will contrast favourably with that of any other city in the United Kingdom when pure water shall be supplied to every house, when sewers are extended, and a few other improvements effected without increased taxation. A subject upon which," he adds, "our citizens are just now rather sensitive."*

I am afraid that this sensitiveness has now become chronic.

The paper from which I have quoted was written in 1865.

The Corporation did not then possess the vastly enlarged powers conferred on them by the Sanitary Act of 1866. An abundant supply of pure water has since been procured for our city; many miles of sewerage have been constructed at a cost of over £100,000; and we are paying one shilling in the pound more taxes than in 1865, and yet Dr. Mapother's predictions have scarcely been fulfilled.

The death-rate has not decreased. It was 26 per 1,000 in 1865. The epidemic of cholera in the following year raised it to 29 per 1,000.

In 1871 the rate was again 26 per 1,000, and last year another terrible epidemic† fell on us apparently with equal force, and *again* it rose to 29 per 1,000.

This death-rate is much higher than we might reasonably expect having regard to the situation of Dublin, the density of the population, and other circumstances; and, moreover, it is pretty certain that the actual death-rate is in excess of

---

* The Act referred to by Dr. Mapother, 27 & 28 Vict., cap. cccv., also provides (sec. 12) that when the Corporation shall have provided a slaughter-house or slaughter-houses for the use of the borough, they may purchase the licences of existing slaughter-houses, &c.
Has any use been made of this power?
† Dr. Grimshaw states that the cases of cholera, during the epidemic of 1866, within the Dublin Registration District, numbered about 2,500 with 1,186 deaths; while during the recent epidemic of small-pox there were about 12,000 or 15,000 cases, and 1,647 deaths!—*Dublin Journal of Medical Science, July*, 1873.

the rates I have quoted. All the deaths which occur are not registered. I shall mention one curious fact as affording some ground for this assertion :—

In 1871 the number of interments *in the three Dublin cemeteries* was 8,536. The total number of deaths registered in the entire Dublin Registration District for the same period was 8,144; in other words, the number of burials exceeded the number of registered deaths by 392. The excess in the previous year was 552.*

Again, in 1865 the birth-rate was 28 per 1,000, and the death-rate 26.

In 1872 the birth-rate was 27 per 1,000, and the death-rate 29.

For the same year in Glasgow, which has the reputation of being a very unhealthy city, the birth-rate exceeded the death-rate by 13 per 1,000, whereas in Dublin the death-rate exceeded the birth-rate by 2 per 1,000.†

Dr. Mapother speaks of the improvements which were to be effected in the tenement houses of the city.

My friend Mr. Henderson in his lecture last Saturday referred to the state of these houses, and regretted that there was no law to compel the owners of such houses to provide proper sanitary arrangements for them.

Let me tell you there is such a law; it is to be found in the Sanitary Act, 1866, sec. 35. I have here a copy of " Regulations for houses and parts of houses let in lodgings, *and occupied by members of more than one family,* within the borough of Dublin," issued in pursuance of the Act of Parliament by the Public Health Committee, and bearing date December 4, 1866.

These regulations provide that every person must have 300 cubic feet of air. That every tenement house must have a properly constructed ashpit and other sanitary accommodation, and these must be kept cleansed and in proper repair. The roof and walls are to be kept in *reasonable* repair and impervious to wet; the windows must open, and the yards, halls, staircases, passages, and rooms not *papered* or *oil-painted* must be lime-washed at least every six months.

There is also a regulation forbidding any person to throw

* The number of burials in the graveyards in the city of Dublin, and cemeteries in its immediate vicinity in 1860 was 8,099. The population of the city was then 254,808. It is now only 246,326, or about 8,482 less than in 1860; while the number of interments is greater by 437! The number of deaths registered in the Dublin District in 1872 was 8,973—the number of interments was 10,056— showing an excess of interments over *registered* deaths of 1,083.

† The birth-rate in London was 35, and Edinburgh 32. The death-rates were respectively 21 and 26 per 1,000.

refuse, &c., in any street or place other than that provided for the proper deposit thereof.

What more do we want than these regulations, *if they were only enforced?*

But are they enforced? A visit to any of the poorer parts of the city will supply you with convincing proof that they are not.

You will not have very far to go in order to find specimens of these miserable tenement dwellings. They abound everywhere, in fact they form more than one-third of the total number of houses in Dublin (*i.e.,* 25,042).

I could point out to you many a locality to which Charles Dickens' description applies with perfect accuracy :—

" It is a black dilapidated street, avoided by all decent people ; where the crazy houses were seized upon when their decay was far advanced, by some bold vagrants, who, after establishing their own possession, took to letting them out in lodgings. *Now,* these tumbling tenements contain by night a swarm of misery ; as on the ruined human wretch vermin parasites appear, so these ruined shelters have bred a crowd of foul existence that crawls in and out of gaps in walls and boards, and coils itself to sleep in maggot numbers where the rain drips in ; and comes and goes, fetching and carrying fever, and sowing more evil in its every footprint than all the fine gentlemen in office shall set right in 500 years, though born expressly to do it."*

Look in at the doorway of one of these wretched abodes. The walls are black with dirt. There is not a trace of the lime whitening enjoined by the regulations. In fact white would seem to be an impossible colour in such an atmosphere. A glance at the shattered roof will be sufficient to show that it is not " impervious to wet ;" and surely that filthy passage is not the place provided "for the proper deposit of the house refuse," which covers the floor to the depth of several inches—and yet what you see from the outside indicates but very feebly the misery, moral and physical, within. How much of that misery is directly due to *our* neglect and apathy, I shall not stop to inquire. I suggest it as a subject for your reflection.

I should like to know what attempt was made to enforce these regulations in two houses (Nos. 17 and 18) in Great

* "The Sanitary Acts are only permissive, and partial in their administration. Owners of house property defy interference, and the authorities are supine. By a righteous retribution the expenses saved by non-administration of the law are consumed by surgeons' bills, extra relief, and the cost of paupers."

"Englishmen and Christian men tolerate scenes of festering corruption for both soul and body where everything tends to crush self-respect, and engender and facilitate vice."—*Archdeacon Sandford.*

Ship-street, in which my friend Dr. J. W. Moore found a population of eighty-three souls, of whom six were in fever; or in those now celebrated fever dens, No. 2, John-street, and No. 4, Mullinahack, with their population of seventeen families, and which furnished sixteen cases of fever to Cork-street Hospital within two years.

These I can assure you are not isolated instances. I sincerely wish they were.

Now, surely the law is not to blame for the state of these tenement-houses. The fault must lie with those who administer it.

Mr. Norwood, who is a valuable member of the Public Health Committee, recently stated, in a paper read before the Statistical and Social Inquiry Society, that there are in this city 9,300 tenement houses, each house containing, on the average, eleven persons.

About one-third of these houses require constant sanitary supervision, another third are not in quite so bad a state, while the remaining third only need occasional inspection.*

Now, the entire sanitary staff employed by the Public Health Committee (exclusive of the Medical Officer of Health, City Analyst, and Secretary) consist of 14 men, viz:—

> 2 Superintendents,
> 8 Sanitary Sergeants,
> 4 Constables,

who are all members of the Metropolitan Police Force, and are under the control of the Commissioner of Police, *as well as* of the Corporation.

How could these 14 policemen efficiently carry on the inspection of the 9,300 tenement houses, and at the same time discharge all their other sanitary duties ? The thing is physically impossible.

We must not blame the officers, but the Public Health Committee, who persistently refuse to provide a sufficient number of *properly qualified* inspectors.†

In Glasgow, with a population only double that of Dublin, things are very differently managed. Here the sanitary arrangements are under the control of a Chief Medical Officer

---

* Mr. Norwood mentions that about 1,000 of these houses belong to three individuals!

† In a statement furnished to the Dublin Sanitary Association in February last, the Public Health Committee declare that " the duties suggested to be undertaken by the dispensary physicians, as district officers of health, are already satisfactorily discharged by eight sanitary sergeants, and two superintendents."

of Health with a competent salary; who has to devote his entire time to the duties of his post.*

Then there are district medical officers, a chief *sanitary* inspector, district *sanitary* inspectors, *nuisance* inspectors, *lodging-house* inspectors, *epidemic* inspectors, &c.

The instructions prepared for these various officers are well worth the perusal of anyone who takes an interest in the subject. I wish that our corporators could be induced to study them, as I think they might then take a higher view than they do now, of the duties and qualifications of sanitary inspectors, as distinguished from searchers for nuisances. At present the Public Health Committee seem to think that a sanitary officer must of necessity be a police-man, and they assign as a reason for not appointing additional inspectors, that the Commissioner of Police cannot spare any more members of the force !†

To take another illustration—There are rules in force in this city for the periodical removal of "manure or other refuse matter." ‡

But no one pays the smallest attention to them.

I wonder how many of the citizens are aware that "all manure, or other refuse matter," must be deodorized, disin-fected, and removed beyond the city boundary, from stables, slaughter-houses, and other like places every day before 7 o'clock, A.M.

The "proclamation" to which I am referring, concludes thus:—" If any of the terms of this *announcement* be disre-garded, immediate proceedings will be taken against the parties in default *without further notice:* for the purpose of enforcing the penalties provided by the statute."

The penalty provided by the statute is 20s. a day; it is not a penalty *not exceeding* 20s., but 20s., neither more nor less.§

There is no clumsy procedure required here. Mr. Byrne in his edition of the Sanitary Acts observes that‖ "One notice served by the authority upon parties for the periodical removal of manure, is sufficient to call its powers into action." In my opinion the publication of the " proclama-

* The salary of the Medical Officer of Health of the Borough of Dublin is £150 per annum.
† In the statement referred to on page 30 the Public Health Committee repeat that in their opinion the duties which the Sanitary Association recommended should be intrusted to medical men, would be most efficiently discharged by policemen,' owing to " their training and the facilities afforded them as constables."
‡ Notice issued by Public Health Committee under 29 & 30 Vic., cap. 90, sec. 53, Sept. 1, 1871.
§ Sanitary Act, 1866, (29 & 30 Vic., cap. 90,) sec. 53.
‖ Compendium of Irish Sanitary Law, p. 45.

tion " which I have quoted, is sufficient to render every
person who does not comply with its directions liable to the
penalty of £1 per diem, without any further notice.

Why do not the authorities set a good example by removing
those frightful nuisances which exist in the Corporation
depôts ?

It is absurd to expect that the public will obey the laws
which they themselves utterly disregard ?

Anything better calculated to bring the laws into contempt
than this practice of issuing regulations, and then not enforcing them, can scarcely be imagined.

But it is perfectly plain that these regulations could not
be enforced and therefore ought never to have been made.
They are impracticable. But why? Because the Corporation will not do, or at any rate has not done what the
nuisance authority ought to do, before it makes such rules,
render it possible for the citizens to obey them, by providing
some means for the removal of manure, house refuse, &c.
This has been done in Glasgow, Manchester,* Liverpool, and
other towns, in which there is a proper sanitary organization,
and it *must* be done in Dublin before long.

I will mention only one other instance of the way in which
the health of this city is protected.

Notwithstanding the experience gained in the late epidemic
of small-pox, the Public Health Committee have provided
no *proper* conveyances for removing such persons suffering
from infectious disorders to hospital. It is true there are
some cabs attached to the hospitals and workhouses,† and
there are two cabs provided by the Public Health Committee.

But cabs are not *proper* conveyances. We want *ambu-
lances* constructed on approved scientific principles. Dr.
Mapother has condemned the employment of cabs, on the
ground that " the change from the recumbent, to the sitting
posture, is most hurtful to the patient.‡ Indeed this is a
subject upon which most medical men are agreed.

The attention of the Corporation has repeatedly been
called to this defect; but they seem to consider the existing
arrangements perfect.

* In Manchester the net amount spent on the removal of night soil, &c., last
year was over £12,000; any private occupier can have his ashpit promptly cleansed
by sending notice to the proper office. Ashpits of tenement houses, &c., are
periodically emptied, and all this is done free of charge. See 2nd Report of Dublin
Sanitary Association.

† Dr. Benson Baker in an article in the British Medical Journal informs us that
he found only one fever cab connected with the Dublin hospitals. This was in
September, 1871. On one occasion he saw a small-pox patient drawn to hospita
on a greengrocer's barrow followed by a crowd of sorrowing friends.

‡ Statistical and Social Inquiry Society Journal, Vol. 2, iv., p. 207.

They certainly can plead that sanitary improvement would lead to an increase in the taxation of the city which even now is enormous; but it might be found that if economy were practised in some of the city departments, a larger proportion of the civic revenues could be made available for sanitary purposes than is the case at present.

I have dwelt at some length on the administration of the sanitary laws in this city, because I am more familiar with the working of the system in Dublin, than elsewhere. I have no wish to lead you to believe that the Public Health Committee of the Corporation of Dublin is less efficient than other similar bodies throughout Ireland. On the contrary, I dare say that in point of efficiency, the committee deserves the first place among our local authorities.

I am well aware of the enormous difficulty of the work which the committee has undertaken, and I am glad to bear my humble testimony to the valuable services it has rendered to this city. At the same time I suppose that hardly anyone will be found to contend that the sanitary condition of Dublin is what it ought to be, or what under a more vigorous administration, it might be.

Now one of the most important problems connected with sanitary legislation is—how to secure that the sanitary laws shall be administered with vigour and efficiency. Various solutions of this problem have been proposed.

An idea seems to be gaining prevalence, that local unpaid bodies, such as town councils and town commissioners, are not fit to be trusted with the care of the public health. These bodies certainly do not appear to entertain a proper sense of the responsibilities of their position. Surely if they did they would bestow a little more care upon the performance of their legitimate duties, and spend less time in discussing politics, and other matters with which they have nothing to do.

A noble Lord once remarked of a certain town council in this country, that its members seemed to think that their mission was to preserve the balance of power in Europe; to maintain the integrity of the Ottoman Porte; and to exercise a general control over the foreign and domestic policy of the government.

Now-a-days "politics are more attractive to *orators*, and more exciting to constituents, than the health and well-being of the people, which the law-givers of ancient empires made their chiefest care."

In a paper read before the Social Science Congress at Plymouth, Mr. Bulteel, a surgeon of large experience in that

o

place, draws a deplorable picture of the degenerate condition
of these local bodies.

He says—

" Within the last twenty years, the materials of which our town
councils and local boards consist have sadly deteriorated ; and
in many places matters have come to such a pass, that it is with
the greatest difficulty a private gentleman, or a first class trades-
man, can be induced to offer himself as a candidate for municipal
honours, because he feels that amidst the storms, personalities, and
jobberies, of local bodies, his position would be unpalatable and
untenable."

Many persons think that town councils and local boards
should be retained for ornamental purposes only, and that
their duties and powers should be transferred to paid, and
responsible officials.

I am afraid that this idea is Utopian. In the first place
the public would not consent to such a revolutionary pro-
posal.* And again, on political grounds, such a change
might be unwise.

The Sanitary Commissioners (1869) observe that—

" The principle of local self-government is of the essence of our
national vigour.

" Local administration, under central superintendence, is the
distinguishing feature of our Government.

" The *theory* is that all that can be done, should be done by the
local authority, and that public expenditure should be controlled
by those who contribute to it."

Assuming then, as I think we may, that our present
system of local administration must be retained, it remains
for sanitary reformers to try and counteract the known evils
of that system.

The public must aid the efforts of the sanitary reformers
by taking more interest than they do at present in the
election of the local bodies. Surely it is due to the apathy, and
indifference of the respectable portion of the community,
that our town councils have been allowed to become what
they are. No one will pretend that the wealth and intelligence of this city are adequately represented in our Corporation. I venture to suggest to the members of the Sanitary
Association, that they should try and remedy this evil by
securing the return of men fitted by position and education
for the discharge of the important and responsible duties,
which they are called upon to perform.

* In New York the sanitary administration of the city has been taken out of the
hands of the Corporation, and vested in paid commissioners nominated by the
government of the state. The plan has proved successful, and is being followed in
other states.

The Sanitary Commissioners have expressed themselves on this point in very clear and forcible language. They express their profound conviction—

"That no code of laws however complete in theory, can be expected to attain its object, unless men of superior education and intelligence feel it their duty to come forward to take part in its working.

"The governing bodies must possess a fair proportion of enlightened and well informed minds. A more vigorous and intelligent public opinion on sanitary matters has yet to be created in many places, and until it is created, the action of the authorities will be more or less hesitating and inconsistent. So large a discretion must be left to local authorities as to details, that in practice, much will always depend on the energy and wisdom, of those who compose such authorities.

"It seems therefore peculiarly incumbent on all who have leisure, to take their share in administering the laws. Their labours may be crowned with little honour, and will be rewarded with no emolument, but if they should hold out small temptation to ambition, there are higher motives for them in public spirit and a sense of duty.

"No institutions of voluntary benevolence are more popular, or more efficiently administered, than hospitals.

"Not only money, but time, and a large share of personal superintendence is given by their supporters; and it may fairly be asked whether to prevent disease—(at any rate to endeavour to prevent it)*—is not as worthy an object as to remove it, and whether it can be better prevented, than by giving full effect to the laws enacted for that purpose."

But it may be asked how can we secure the presence of men possessing a scientific knowledge of sanitary matters, on our local boards?

This is certainly a difficulty.

The Legislature wisely conferred a power on the various local bodies, of delegating their functions to a committee comprised *wholly* of their own members, or *partly* of their own members and *partly* of ratepayers not members of the Board.†

I do not know of a single instance in which this power of co-opting strangers has been exercised. It fully meets the difficulty I have alluded to.

I have reason to believe that in our own city the Corporation would be able to avail themselves of the learning, and great practical experience, of some of our most distinguished

* Sir W. Jenner says, "I have always taught that the highest branch of medicine is '*preventive*' medicine."
† San. Act, 1866, sec. 4.

medical men, who would be willing to make some sacrifice
of their time and convenience for the public good.

But the truth is that the Corporation do not desire to be
interfered with, by strangers to their own body.

They have recently excluded some of our leading citizens
from the Gas Committee, on the ground that they were able
to do their own business without help from outside. Possibly
if they had courted the assistance of the citizens, the fate of
the Gas Bill might have been different.

After this public profession of faith in their own adminis-
trative capacity, it is very improbable that the assistance of
outsiders will be sought in carrying out the sanitary laws,
unless, indeed, under the strong pressure of public opinion,
and I fear that public opinion is not yet sufficiently enlightened, to exert such a pressure.

But the question remains, how are we to render the
present system effective?

It seems to me that the remedy lies in the appointment
of skilled and *specially qualified* medical officers of health,
to superintend the sanitary administration within their
respective districts.

Mr. Michael, a member of the English bar, who has
devoted a large share of his attention to sanitary matters,
emphatically asserts that—

" No system can possibly succeed which trusts for its execution
to unpaid members of Boards, instead of throwing the onus both of
action, and responsibility, upon competent and competently paid
officials.

" A gentleman of special training, having devoted special study
to hygiene, and relieved from the cares of private practice, would
work incalculable good to any district which enjoyed the benefit
of his services."

The Sanitary Commissioners did not consider it necessary
that the health officer should be debarred from private
practice ; but so far as I am able to judge, the balance of
opinion, at least in England, amongst persons competent to
pronounce on the subject, is largely in favour of such a
restriction, at least as regards " chief " officers of health.

The " Joint Committee," to which I have referred, in their
report for 1871,* state that—

" Their estimate of the functions to be discharged by the medical
officer of health ; of his special qualifications, and of the time to be
devoted to the discharge of his duties, contemplates a class of
officers entirely special, and without the distractions and difficulties, which ordinary practice would necessarily entail."

In an admirable pamphlet* on this subject by Mr. Ceely of Aylesbury, the author urges the necessity for having two orders of health officers—the one, engaged in practice, as union medical officers, and allowed to attend private patients; the other, debarred from general practice, receiving reports from the former, and acting over counties, or first-class boroughs.

Mr. Ceely appends three minutes of the Board of Health, which existed from 1848 to 1858, and they certainly prove that the members of the Board took a more exalted view of the province of preventive medicine, and the duties and qualifications of the health officer than now prevails.

These documents afford to my mind unanswerable reasons for the appointment of chief health officers, who shall give their time exclusively to their official duties.

The case of Liverpool may be mentioned, "where the then Home Secretary refused to sanction the admixture of private practice, with public duties, and subsequent experience has shown the soundness of that decision."

The difficulty at once suggests itself, that every sanitary authority throughout the country cannot afford to maintain officers of the stamp, I have indicated.

Various solutions of this difficulty have been proposed. Extended sanitary areas, county Boards, &c.

I fear, however, that none of the plans suggested for England, would be exactly suited to this country.

I venture to suggest the following, as a way in which the difficulty might be met.

I have already shown that the number of sanitary districts in Ireland might easily be reduced to 207—i.e., 44 urban, and 163 rural.

Let each of these districts have its own medical officer, who, for obvious reasons, must be the local dispensary physician.

These officers should be allowed to pursue their private practice, and a comparatively small remuneration would compensate them for their increased duties.†

The whole country might be divided into large districts, each under the control of a chief medical officer of health,

* "Health Officers—their Appointment, Duties, and Qualifications." London: Richards, 37, Great Queen-street.
† Dr. Toler Maunsell, Honorary Secretary of the Irish Poor Law Medical Officers' Association, has just published a very useful analysis of the population, salaries of medical officers, &c., &c., of the various counties, and rural, and urban districts in Ireland, from which it appears that there are 800 dispensary medical officers. Their average salary is under £100 a year! Their salaries represent an annual poundage of ¾d. on the valuation. The number of patients attended during the year was 741,275.

who should reside in a central part of the district, and be in
constant communication with all the local officers.

This chief officer should be debarred from private practice,
and should possess the special qualifications required for the
post.

He should be the servant of the central authority, and be
entirely independent of all local interests.

If his directions are not carried out by the local authority
the default should at once be reported, and the necessary
steps taken by the central authority to compel action.* Of
course it would be part of his duty, to make frequent inspections of every town within his district.

The extent of the district to be allotted to each of these
officers, is a matter which could easily be determined by the
central authority.

It seems quite possible that the work could be efficiently
carried on by eight chief officers, who might be located, say,
at Enniskillen, Belfast, Dundalk, Dublin, Galway, Limerick,
Waterford, and Cork.

Each district would include, on the average, about twenty
unions, containing a town population of less than 190,000,
and a rural population of about 490,000.

In some such way as this an efficient system of sanitary
inspection and control might be carried out, sanitary action
stimulated, and some beneficial results be reasonably expected to follow.

The expense of such a system would be small.

The salaries of the chief officers would range from £1,000
to £400 a year each, and need not exceed in all £5,000
a year. The State should pay these officers and their
staffs. It bears half the expenses of the Poor Law medical
officers' salaries, and ought not to grudge a few paltry thousands for the promotion of public health. "The result of
local neglect would be national mischief, so the prevention
of such a result is matter for State interference, and to be
purchased, at least in part, at the State expense." The mere
money cost of public ill-health owing to the loss of labour of
the sick, and of those who attend them, and by increased expenditure, must be estimated at many millions a year.†

* At present the duty of reporting the neglect of the local authority rests with
the public (under sec. 49 of the Sanitary Act, 1866), who seem unwilling to avail
themselves of the privilege.

† The number of those who were affected by the late small-pox epidemic in this
city, and who (together with their friends) were obliged to apply to the Mansion-
house Relief Committee was 6,000.—*Dr. Grimshaw's Report.*

"In a money point of view few things have ever paid better than the outlay which
Bristol has made, in the appointment of a Health Officer, assisted by an active staff,
for the repression of disease among her citizens."—*Dr. Budd.*

But we want something more than good laws and active administration. To quote again from Mr. Michael—

" To be thoroughly successful, sanitary action must prevail within, as well as without, the dwelling. This is beyond the power of any law which a man does not enact and execute in his own house.

" The full benefit of personal action can never be experienced, until such a knowledge of sanitary science, and such a belief in its efficacy prevails, as shall constitute every man and every woman in his, or her own household, an officer of health.

" To secure such action we want a better condition of education in its practical aspects, for without this, sanitary legislation can never secure those triumphs which it is calculated to achieve."

This is true of nearly all social reforms. The masses must be taught to appreciate them. At present they are unable to understand the object of our efforts. They have been brought up in dirt, and lived in dirt, and never knew that they were anything the worse of it, and naturally enough they object to our interference with their fancied comforts. As Dr. Lankester has well observed—

" It is the ignorance of the poor, this want of knowledge of sanitary laws, a knowledge of the laws of life, the laws by which God governs the life of the community, their ignorance of the *value* of fresh air, the *value* of pure water, the *value* of warmth, the *value* of many things which they could use and employ, which causes them to die."

The only means of getting rid of this popular ignorance is education.

Let the people, rich and poor, once understand the importance of obeying the laws of health, and the difficulties of sanitary administration will be greatly reduced. The public will see the necessity for what they now regard as absurd and unnecessary regulations, and will be far more willing to conform to them.*

The common principles of sanitary science should be taught in all our schools, public and private, and then there might

---

* In Glasgow I find that provision is made by the authorities for the appointment of female visitors, whose duty it is to instruct the poorer classes as to cleanliness of person, cleanliness of houses, &c., and to point out how they and their children may imitate and acquire in these respects, the habits of the better classes. The visitors are cautioned not to spend time in frivolous gossip.

In the first report of the National Health Society just published it is suggested that " ladies might gather round them mothers, working women, and girls, and read or talk to them of some of the many subjects of health, and daily life, on which knowledge would be valuable ; in a simple, and easy, but truthful manner."

be some chance, that the coming race would appreciate cleanliness a little more than their parents appear to do.*

At present so far from the principles of health being taught in our schools, they are generally conducted in defiance of those principles.

Mr. Edwin Chadwick, C.B., gives it as his opinion that schools are—

"The centres of children's epidemics, and that an excess of 50,000 deaths annually in England and Wales, in the school periods of life, is due to the massing of unwashed children in ill-warmed, and ill-ventilated schools, and keeping them for hours together under these conditions."

The way to remedy this state of things is obvious. Let schools be provided with proper baths, and lavatories, and let them be properly heated, and ventilated. The pupils will then be taught habits of cleanliness, and the school will cease to be, what it too often is now-a-days, a centre of infectious disease.

In selecting a plan for a school building the sanitary arrangements should first be looked to. The architectural design is really of very secondary importance.†

The education of the adult portion of the community may be accomplished by means of lectures and classes, and by the dissemination of pamphlets on sanitary subjects.

This education must not be confined to the humbler classes of society, for I believe that a disregard of the laws of health is by no means peculiar to those classes.

There are plenty of law breakers to be found amongst the upper classes, whose mansions the sanitary inspector would not dare to visit, but where his services are often sadly needed.

---

* In an article on "The Science of Health" in "Good Words" for January last, Canon Kingsley advocates the establishment of public schools of health in every large town in the kingdom.

I have recently heard that a large sum of money has been given to the Birmingham and Midland Counties Institute to found a lectureship on the laws of health. The lectures are to be chiefly for the benefit of the working classes of both sexes.

† Speaking of badly ventilated schools I may observe that the Sanitary Commissioners of 1843 recommend, "that measures should be adopted for promoting a proper system of ventilation in all edifices intended for public assemblage and resort, especially these for the education of youth." It were time that this recommendation which has hitherto been completely ignored by the Legislature, should be carried into effect, for I feel sure that a vast injury to public health must result from the utter absence of proper ventilation, in nearly all our churches, court-houses, theatres, and other places of assembly.

‡An admirable series of sanitary tracts has been issued by the Ladies' Sanitary Association of London.

We cannot expect any very important Sanitary reforms until public opinion is strong enough to insist on them. It must be borne in mind that all social reforms are difficult of accomplishment, and we are not to be discouraged if Sanitary reform proves no exception to the rule.

As education advances and as public opinion becomes more enlightened, we may hope to see reforms effected which now appear impracticable.

It is our duty to aid in the formation of such a sound public opinion, by endeavouring to spread a practical knowledge of the laws of health among all classes of the community. The public mind needs to be aroused to a sense of the vast importance of the subject. "The public health" has been too long neglected by the legislature and by society at large; but I believe that the time is not very far distant when it must receive from both the attention which it deserves.

One of the objects of the "Sanitary Association," which has recently been established in this city, is "The creation of an educated public opinion with regard to Sanitary matters in general." I hope that before many months are past, we may have associations labouring for the same object in every large town in Ireland.*

Several organizations for promoting a knowledge of Sanitary science are at work in England, amongst which I may mention "The National Health Society," and "The Ladies' Sanitary Association." The latter of these societies has branches in various parts of England and Scotland. Voluntary organizations such as these will materially assist in educating public opinion, and so prepare the way for legislation, besides doing much to strengthen the hands of the authorities.

Some persons seem to regard Sanitary Reform as if the sole aim and object of its promoters were simply to effect some trifling reduction in our annual death-rates. A proper system of Sanitary administration would no doubt have this result, and probably to an extent which the public would scarcely be disposed to credit.

But this is not all we hope to effect. We seek not merely to rescue a few victims from the grave to which neglect would consign them, but also to deliver tens of thousands of our brothers and sisters from the physical suffering and from the moral pollution which necessarily follow from the deplorable condition of the masses throughout the country,

* Since this lecture was delivered a Sanitary Association has been formed in Cork.

P

chiefly amongst our town populations. As a general rule the dwellings of the poor in great towns are a disgrace to our boasted civilization. The frightful evils, physical and moral, which result from the overcrowding of large numbers of people without regard to age, or sex, in wretched, pestilential hovels are beginning to make themselves felt.

The authorities seem unable or unwilling to grapple with this difficulty, but I am glad to say that, thanks to the philanthropic efforts of a few individuals, societies have been formed in various parts of the United Kingdom, for improving the dwellings of the poor.

One of these societies, "The Artisans', Labourers' and General Dwellings Company (London)," has a capital of over £50,000, and pays its shareholders a dividend of £6 per cent. The directors state that in the houses erected by them, the average death-rate has been 6 per 1,000, while in neighbouring localities it exceeded 25 per 1,000.

"The Improved Industrial Dwellings Company" (London), has invested half a million of money, and pays a guaranteed dividend of £5 per cent. It affords accommodation to about 9,000 persons.

Some charitable persons in London have done much good by purchasing houses in poor neighbourhoods, putting them into substantial repair, and then letting them in lodgings, subject to stringent Sanitary regulations (which are enforced). This seems a more sensible plan, than that of building large mansions at great expense, and fitting them up in a costly manner—in fact, making them unsuitable for the purpose for which they were intended.

In Glasgow an enormous improvement has been effected by the Corporation, under the powers of a special Act of Parliament. Large districts have been purchased from time to time, uninhabitable houses have been demolished, and new ones erected on their sites, at a *comparatively* small cost, provided by local taxation. We want something of this sort in Dublin, where there are over a thousand houses unfit for habitation, and which require demolition. The powers conferred on the Corporation by the "Artisans' and Labourers' Dwellings Act," to which I have referred at some length, might be found sufficient for the purpose. If not, additional powers should be sought from Parliament.*

If the Corporation neglect their duty in this respect, that

---

* Overcrowding from want of space for building creates special diseases, and completely demoralizes the people. If I were to pitch upon one thing which is the cause of the epidemic disease, and physical and moral degradation of the population. I should say it was the system of house construction such as we have had in Glasgow for three or four generations.—*Dr. Gairdner.*

is no reason why we should neglect ours. A good example has been shown us in England, and there is no practical difficulty in the way of our following it here.

Whatever we do let us not commit the fatal mistake which I believe has done much to retard the progress of this country—that of regarding "everything as good enough, well enough, time enough."

The Sanitary movement has, as I have said, been inaugurated in Dublin. The success of that movement must depend altogether on the support, material and moral, which you, the public, afford it. I feel sure that no one who feels an interest in the well being of his race, can refuse to take part in this work.

I do not care to appeal to any of the selfish considerations which might be suggested as inducements to exertion, such as the cost of sickness, and the pecuniary saving to be effected by its prevention, or the still more frequently urged plea, that we cannot neglect the health of our neighbours without risk to the health—it may be the lives—of ourselves and our families. I prefer to rest our claims to your active sympathy, on those broad principles of Christian charity, which are common to all our creeds.

FINIS.

DUBLIN: Printed by ALEXANDER THOM, 87 & 88, Abbey-street.
For Her Majesty's Stationery Office.
[753.—150.—1/74.]

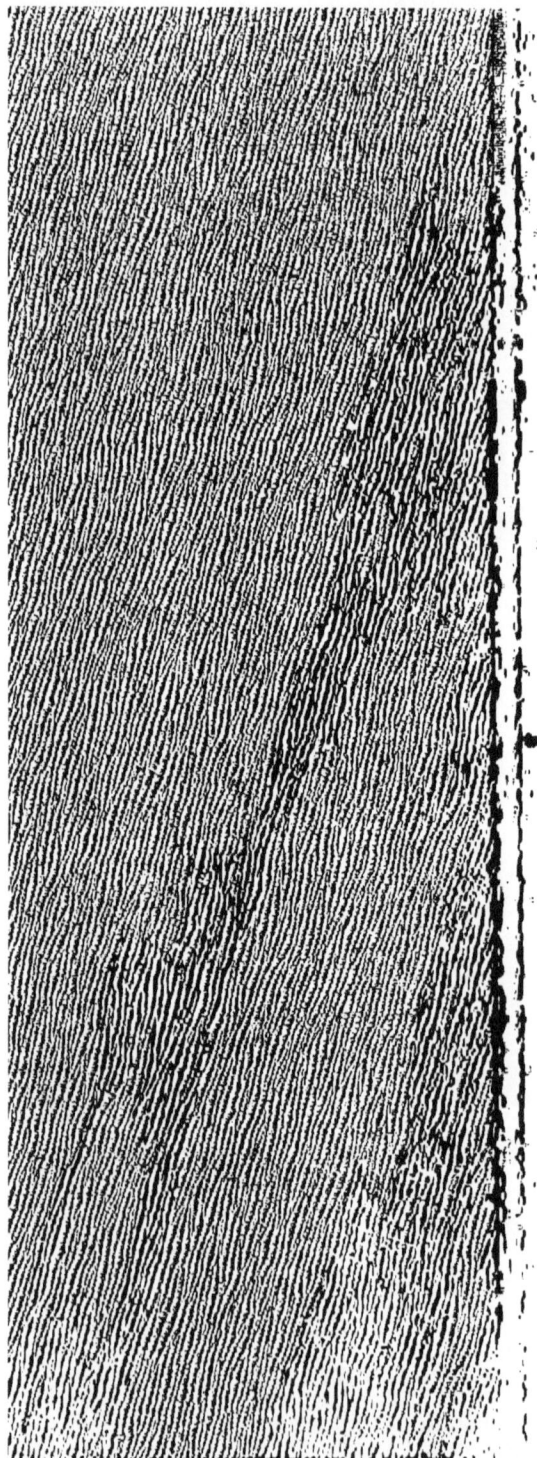

www.ingramcontent.com/pod-product-compliance
Lightning Source LLC
Chambersburg PA
CBHW021656210326
41599CB00013B/1443